Mathematics for Agriculture

Second Edition

Mathematics for Agriculture

Revised by

Betty C. Rogers

Associate Professor of Mathematics
Piedmont College
Demorest, Georgia

Original Edition by

Clifford M. Hokanson

Former Assistant Professor
Related Education Division
University of Minnesota Technical College–Waseca

Interstate Publishers, Inc.
Danville, Illinois

Mathematics for Agriculture

Second Edition

First Edition published under the title
Applied Problems in Mathematics for Agriculture, 1984

ISBN 0-8134-3174-3

Order from

INTERSTATE PUBLISHERS, INC.

510 North Vermilion Street
P.O. Box 50
Danville, IL 61834-0050

Phone: (800) 843-4774
Fax: (217) 446-9706
Email: info-ipp@IPPINC.com

World Wide Web: http://www.IPPINC.com

PREFACE

Production agriculture and its support industries produce a large portion of the United States gross national product. Increasing technology used in the growing, processing, and distribution of food and fiber make it mandatory that workers in agricultural occupations have skills in the analysis and solution of mathematical problems.

The primary intent of this text is to provide an agriculturally relevant review of basic arithmetic, statistical interpretation, and algebraic concepts to prepare the student for the mathematics involved in other courses. However, the text would readily lend itself for use as a general math text in an agricultural curriculum.

The application problems use current and realistic agricultural situations similar to those encountered by professionals engaged in production agriculture, the raising of crops and livestock, and by those employed in agriculturally related occupations and industries. Because the math skills required in solving problems in the text range from the very simple to the more challenging, the text could be adapted for use in middle school through high school, in two-year college programs, or in college developmental algebra courses.

I wish to acknowledge the extensive use of Internet sites of the National Agricultural Statistics Service and the USDA Economics and Statistics System as sources of the current and realistic agricultural statistics used in the text.

I wish to thank my family for their patience and understanding during the preparation of this text and Dr. Jasper S. Lee for his encouragement and use of materials from Interstate's "AgriScience and Technology Series," of which he is editor.

Betty C. Rogers

ABOUT THE AUTHORS

Betty C. Rogers is Associate Professor of Mathematics at Piedmont College in Demorest, Georgia. She has degrees in both Mathematics and Higher Education. Dr. Rogers has taught in colleges and high schools for over 20 years and has a Georgia Lifetime Masters Teaching certificate. She grew up on farms in Mississippi and Alabama and owns a farm in north Georgia, where she raises beef cattle and vegetables. In addition to teaching, she facilitates seminars and workshops on group processes, cooperative learning, and team building.

Clifford M. Hokanson has taught in the public schools in Minnesota, has taught in an adult vocational agricultural program, and has had personal experience as a crop and livestock farmer. He was formerly an assistant professor teaching mathematics at the University of Minnesota Technical College, Waseca, Minnesota. UMW is a single-mission agriculturally oriented college, training students for entry into mid-management positions in agriculture and its related occupations.

TO THE STUDENT

Math is all around you—in everything you do and wherever you live or work. Expanding your mathematical knowledge and becoming competent in making computations can enrich your personal life and enhance your career potential.

Even more important than the math course itself is your attitude towards math. You will be more successful in this course if you realize that math can be a key to your future. Expertise in computational skills gives you a definite advantage in any career.

This text is designed to give you a review of basic arithmetic, an overview of statistical interpretation, and a brief introduction to algebra. The applied problems cover a wide range of agricultural situations and should increase your appreciation for the extent of mathematical involvement in agriculture.

The discovery of application problems relating to your special area of interest may spark your enthusiasm for mathematics. Mastery of problem-solving techniques can be an interesting and worthwhile challenge.

Remember, also, that the saying "practice makes perfect" is very applicable in mathematics.

TO THE INSTRUCTOR

The blending of verbal skills with mathematical skills in the solution of "applications" presents a challenge to both the student and the instructor. Students can frequently manipulate mathematical operations, but may not be able to apply these principals to real world situations.

Traditionally, students have abhorred word problems and have gained little mastery in problem-solving techniques. Most of the real-life mathematical situations in either personal or occupational contexts are the "story problem" type. The wide range of types of application problems presented in this text should promote student interest and develop confidence and competence in handling verbal problems.

The material in this text can be adapted to a variety of modes of instruction including cooperative learning, traditional lecture, class with laboratory, or group based active learning. The inclusion of a large number of problems, varying in difficulty, allows the instructor to choose the quantity and type of problems best suited to student needs.

For students requiring reassurance in the performance of the basic mathematical operations, the text outlines the rules and procedures for each operation, including solved examples. Each exercise focuses on a basic operation and begins with a limited number of general or non-verbal problems. These are followed by a variety of agriculturally oriented application problems. The more academically mature student can proceed immediately to the application problems.

Problems in the text readily lend themselves to computations using hand-held calculator. It may be advantageous to force a review of basic arithmetic skills by requiring the problems in the first chapter be solved without using a calculator. However, applications in later chapters will benefit from the use of calculators so students can concentrate in the implication of the problem rather than the manipulation of numbers.

With the passage of metrication legislation by the United States Congress in 1975, some industries and governmental agencies immediately began the changeover to metric. Public resistance to metrication has brought it almost to a standstill. Because agricultural research data are often written in metric units, and the United States must function in a global agriculture society, workers in agriculture need to be literate in both systems.

CONTENTS

1

WHOLE NUMBERS

Key Terms

addends
decimal
difference
dividend
divisor
factors

minuend
multiplicand
multiplier
product
quotient
reverse addend

reverse factor
subtrahend
sum
yield per acre

Career Connections

ANIMAL TECHNICIAN

Animal technicians provide feed and water and maintain facilities for the animals in their care. They must be knowledgeable of the nutritional requirements, signs of sickness, and other special needs of the animals. An animal technician usually provides medication for minor physical illnesses and injuries and must determine when to contact a veterinarian for more serious problems.

Agriculture study in community colleges is nearly essential. Many students get college degrees in animal science or a related area. Practical experience in working with animals is also necessary.

Jobs are found wherever animals are important in the lives of people. Most animal technicians work on ranches, in zoos, or in other animal facilities. This job usually involves extensive time outdoors.

A. Our Number System

The number system we use is called a **decimal** system because we have **ten** symbols (called digits) that make up the system. The symbols are 0, 1, 2, 3, 4, 5, 6, 7, 8, and 9. Numbers are made up of the ten symbols or numerals used in a place-value system. The system gives each symbol in a number its own value as well as a value based on its position in the place-value system. The table below gives the word names of each of the positions in the place-value system:

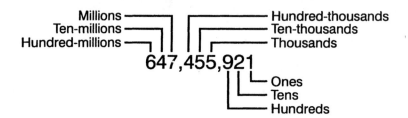

Numbers are usually written with numerals but can also be written in words (as the number would be spoken). The number 98,541,628 would be written as ninety-eight million, five hundred forty-one thousand, six hundred twenty-eight. Notice that we use a comma to group the periods (ones, thousands, millions). Numbers twenty-one through ninety-nine are hyphenated. Later when studying decimals, we will learn that the word **and** is used only to represent a decimal point.

B. Addition of Whole Numbers

The numbers that are to be added are called **addends** and the result of the operation of combining the addends is called the **sum**. In the expression 6 + 2 = 8, the symbol "+" is a plus sign and indicates addition.

Checking Addition

There are a number of ways to check addition for accuracy. One simple method is called the **reverse addend** method. Reversing the order of the addends produces different combinations of addends, thus lessening the chance of repeating errors.

Example:

	The computation		The check
			22,138
	6,234		6,234
Add	8,961	Add	8,961
downward	2,362	upward	2,362
	4,581		4,581
	22,138		

C. Subtraction of Whole Numbers

Subtraction is an operation that determines the **difference** between two numbers or the amount that one number is larger or smaller than the other. To find the difference, subtract the smaller number from the larger number. In the expression, $9 - 5 = 4$, the symbol "$-$" is a minus sign and indicates subtraction. The larger number is called the **minuend** and the smaller number is called the **subtrahend**. The result of the operation of subtraction is called the **difference**.

Checking Subtraction

Adding the difference (answer) and subtrahend (lower number) easily checks subtraction. This sum will equal the minuend (upper number) if the subtraction is correct.

Example:

Computation	Check
869	641
$-$ 228	$+$ 228
641	869

D. Multiplication of Whole Numbers

Multiplication is an operation between two numbers that produces a third number called a **product**. The numbers to be multiplied are sometimes referred to as **factors**. The product of 7 and 3 means that 7 is used as an addend 3 times, that is, $7 \times 3 = 7 + 7 + 7$ (7 added three times). The "7" would be called the **multiplicand** and the "3" is the **multiplier**. In the expression $7 \times 3 = 21$, the symbol "$\times$" is a "times" sign and indicates multiplication; "21" is the product. Other symbols that mean multiplication are a dot placed between the numbers (e.g., $7 \cdot 3$) and parenthesis placed around the numbers [e.g., 7(3), (7)3, and (7)(3)].

Checking Multiplication

One simple method of checking multiplication is the **reverse factor** method. Reversing the multiplicand and the multiplier produces different combinations of factors, decreasing the chance of repeating errors.

Example:

Computation	Check
632	849
$\times$ 849	$\times$ 632
5688	1698
2528	2547
5056	5094
536568	536568

E. Division of Whole Numbers

The operation of division of two numbers results in a third number called the **quotient**. Division may be indicated by the symbol "÷" (divided by) or by a line, called a vinculum, between the numbers such as 6/17. The number to be divided is called the **dividend** and the number by which you divide is the **divisor**. In the expression, 24 ÷ 8 = 3, "24" is the dividend, "8" is the divisor, and "3" is the quotient.

Checking Division

Division is checked by multiplying the quotient (answer) by the divisor. The result should equal the dividend (original number). If 54 ÷ 6 = 9, then 9 × 6 = 54. If there is a remainder in the division, then (divisor × quotient) + remainder = dividend.

Example:

Computation	Check

```
         108              108
     27)2938            ×  27
        27               756
        23               216
        00              2916
       238            +   22
       216             2938
        22
```

F. Tests of Divisibility

1. **A number is exactly divisible by 2 if the ones digit is divisible by 2 (any even number and 0).**

    Examples: 49$\underline{8}$, 4$\underline{0}$, and 29$\underline{2}$ are divisible by 2; 36$\underline{5}$, 28$\underline{1}$, and 5$\underline{7}$ are not.

2. **A number is exactly divisible by 3 if the sum of its digits is exactly divisible by 3.**

    Examples: 453 → 4 + 5 + 3 = 12. 12 is exactly divisible by 3, therefore 453 is divisible by 3.

    772 → 7 + 7 + 2 = 16. 16 is not exactly divisible by 3, therefore 772 is not exactly divisible by 3.

3. **A number is exactly divisible by 4 if the number expressed by the tens and ones digits is exactly divisible by 4.**

    Examples: 5$\underline{16}$ is exactly divisible by 4 because 16 is exactly divisible by 4.

3<u>15</u> is not exactly divisible by 4 because 15 is not exactly divisible by 4.

4. **A number is exactly divisible by 5 if its ones digit is 0 or 5.**

    Examples: 29<u>5</u>, 42<u>0</u>, and 30<u>5</u> are divisible by 5; 49<u>2</u>, 11<u>4</u>, and 29<u>7</u> are not exactly divisible by 5.

5. **A number is exactly divisible by 10 if its ones digit is 0.**

    Examples: 6<u>0</u>, 19<u>0</u>, and 75<u>0</u> are divisible by 10; 6<u>8</u>, 22<u>5</u>, and 80<u>4</u> are not exactly divisible by 10.

G. Order of Mathematical Operations

In a problem with two or more operations and no grouping symbols, work from left to right completing all multiplication and division first. Next, go back and work from left to right completing all addition and subtraction. If there are symbols of grouping, such as parentheses ( ), or brackets [ ], complete the operations inside the symbols first then continue as above. Some students find the memory device, "<u>P</u>lease <u>M</u>y <u>D</u>ear <u>A</u>unt <u>S</u>ally," helpful in remembering the order of operations: <u>P</u>arentheses, <u>M</u>ultiply, <u>D</u>ivide, <u>A</u>dd, and <u>S</u>ubtract.

Example: $25 - (6 \div 2) + 8 \times 3 = 25 - 3 + 24 = 22 + 24 = 46.$

Example: $2\,[(14 + 6 - 5) \div (8-3)] = 2[(20 - 5) \div 5] = 2(15 \div 5) = 2(3) = 6.$

H. Solving Applied Problems

1. Read the problem carefully.

2. Determine exactly what quantity is to be found.

3. Find the relationship between the given data and what is to be found. Write down the relevant given data. Delete any data that are only descriptive and will not be used in the calculations.

4. It is often helpful to draw a sketch of the problem or set up a chart of the information.

5. Plan the sequence of mathematical operations.

6. Perform the mathematical operations and check each computation for accuracy.

7. Compare the answer with the information in the problem. Does the answer appear reasonable and logical?

Example: Combine A has a harvesting capacity of 620 bushels of corn per hour. Combine B has a 495 bushel-per-hour capacity. How many more bushels of corn can Combine A harvest in nine hours of operation?

Given data: Combine A — 620 bu/hr, Combine B — 495 bu/hr, 9 hours of operation

Find: How many more bushels of corn can Combine A harvest in nine hours?

Plan: Find out the total bushels each combine can harvest in nine hours and then find the difference.

Solution: Combine A — 620 bu/hr × 9 hours = 5,580 bushels
Combine B — 495 bu/hr × 9 hours = 4,455 bushels
Difference = 1,125 bushels

Exercise 1-1

ADDITION, SUBTRACTION, MULTIPLICATION, AND DIVISION OF WHOLE NUMBERS

Write the numerals for the following:

1. Six thousand, four hundred sixty-five. _____

2. One hundred thousand, twenty-nine. _____

3. Sixteen thousand, two hundred thirty-eight. _____

4. One thousand, nine. _____

5. Forty-five million, seven hundred twenty thousand, nine. _____

Express the following numerals in words:

6. 99,633 _____

7. 102,021 _____

8. 73,678,009 _____

9. 2,869 _____

10. 142,684 _____

Find the sum of:

11. 12,936 square feet; 4,816 square feet; and 8,209 square feet. _____

12. 76 kilograms, 108 kilograms, and 1,029 kilograms. _____

Find the difference of:

13. 1,265 cubic yards − 936 cubic yards. *329*

14. 1,925 milliliters − 608 milliliters. *1,371*

Find the product of:

15. 247 meters and 648 meters. (Answer will be in square meters.) *160,056*

16. $9,320 × 13. $ *121,160*

Find the quotient of:

17. 5,824 kilograms ÷ 16. *364*

18. 4,528 gallons ÷ 4. *1,132*

Add and check by the reverse addend method:

19.	20.	21.
469	1,934	29,483
239	2,924	2,009
931	6,288	36
128	5,912	9,125
736	8,707	783

Subtract and check by addition:

22.	40,006 Check	23.	92,708 Check	24.	9,762 Check
	− 9,427		− 48,929		− 7,008

Name_____   Date_____

Multiply and check by the reverse factor check:

25. 20,806   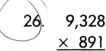 26. 9,328   27. 74,305
     × 4,329   × 891   × 6,466

     Check   Check   Check

Divide and check by multiplication:

28. 4,021)3,761,908   Check

29. 679)9,127,833   Check

30. Use the test of divisibility to answer the following:

    a. Circle the numbers exactly divisible by 2.

            121   246   600   425

    b. Circle the numbers exactly divisible by 3.

            9,372   437   960   712

    c. Circle the numbers exactly divisible by 4.

            812   648   954   1,004

9

d. Circle the numbers exactly divisible by 5.

       368   405   920   731

e. Circle the numbers exactly divisible by 10.

       225   170   608   230

Solve the following problems:

31. A milk hauler picked up the following amounts of milk from three dairy farmers: 4,638 pounds; 6,209 pounds; and 5,206 pounds. What total weight of milk was in the tanker?

                                                              _____

32. A turkey hatchery produced 2,692 poults in January; 3,731 in February; 4,298 in March; and 2,982 in April. How many poults did the hatchery produce in the four-month period?

                                                              _____

33. The first-of-the-month inventory of animals on a dairy farm shows the following information: 1 herd sire, 126 cows, 26 bred heifers, 36 open heifers, 57 dairy steers, and 49 calves. What is the total of the dairy animal inventory?

                                                              _____

34. A 225-kilogram colt is fed a daily ration of timothy hay and a grain concentrate consisting of crimped oats and soybean meal. The hay provides 81 grams of digestible protein (DP); 310 grams of DP are provided by the oats and 145 grams of DP are furnished by the soybean meal. How much DP is the colt fed?

                                                              _____

Name_____ Date _____

35. A truck-garden operator is purchasing a new garden tractor and several attachments. The table below shows the itemized portion of the sales invoice for the equipment. Find the total purchase price.

Quantity	Model	Description	Unit Price		Amount	
1	89GT2145	40 Horsepower Tractor	15,195	00	15,195	00
1	89GT3368	12-inch Plow	600	00	600	00
1	89GT4181	Spring-tooth Harrow	350	00	350	00
1	89GT8173	3-point Cultivator	395	00	395	00

_____

36. A semi-truck uses approximately 2,175 British Thermal Units (BTU) of heat energy per ton-mile to transport grain. If a barge can do the same job using 540 BTU per ton-mile, what is the energy saving on grain shipped by barge?

_____

37. The daily digestible energy requirements of a bred sow during the gestation period is 6,600 kilocalories. Later, when nursing her litter, requirements increase to 18,150 kilocalories per day. What additional amount of digestible energy is needed by the sow during lactation?

_____

38. A vegetable grower harvested and packed 2,776 crates of vegetables. When cooled, 1,917 crates were shipped and the remaining crates put into cold storage. How many crates of vegetables were placed in storage?

_____

39. Last year California's cash receipts (in millions of dollars) for lettuce was 1,084 while Arizona's total was 306. How much more did California sell than Arizona?

_____

40. Field inspection shows six mature pigweed plants per square rod of cropland. If a pigweed plant can produce 117,500 seeds, what is the potential pigweed population per square rod the next cropping season?

_____

41. A veal calf gains 1 pound from approximately 10 pounds of whole milk. What amount of whole milk is consumed by a calf that gains 76 pounds?

_____

42. As a feed for swine, 1 kilogram of rice bran provides 2,706 kilocalories of metabolizable energy (ME). What amount of ME would 3 kilograms of rice bran provide?

_____

43. A landscape designer suggests the use of flexible plastic edging around several flower beds. Measurements indicate six 20-foot rolls of edging are needed. If each roll costs $7.00 what would the edging for the flower beds cost?

_____

44. A certified seed grower raises four varieties of flax. The table below shows the grower's record of the field size and the total yield from each variety. Find the variety having the highest average yield in kilograms per hectare.

Variety	Field Size (Hectares)	Total Yield (Kilograms)	Average Yield (Kilograms/Hectare)
A	12	23,184	1,932
B	9	18,666	2,074
C	11	20,559	1,869
D	14	24,094	1,721

_____

12

45. Figure 1–1 shows the center section of a sprayer boom and the width coverage of each nozzle. If a left and right section of the boom each have the same number of nozzles and nozzle spacing, what is the total coverage (in inches) of the boom?

_____

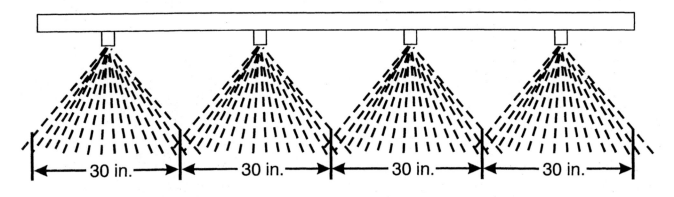

Figure 1–1

46. To avoid excessive spoilage in a large upright silo, a beef feeder needs to remove 1,375 kilograms of silage each day. If each steer is to be fed 11 kilograms of silage per day, how many steers are needed to eat this quantity of silage?

_____

47. A truckload of barley weighs 16,954 pounds. If this amount of barley is sold as 346 bushels, what is the weight for a bushel of barley?

_____

48. The operator of a corn combine, which harvests six 30-inch rows, charged $4,725 for harvesting 189 acres of corn. What was the per acre custom rate charged?

_____

49. A landscape gardener is ordering tulip bulbs to be planted in a circular flower bed. It has been determined that the total area of the flower bed is 56,700 square centimeters and that each bulb should have 324 square centimeters of space. How many bulbs should be ordered?

_____

50. Figure 1–2 shows a tractor pulling a seven-furrow plow that is plowing a strip 112 inches wide. What is the width of each plow bottom?

_____

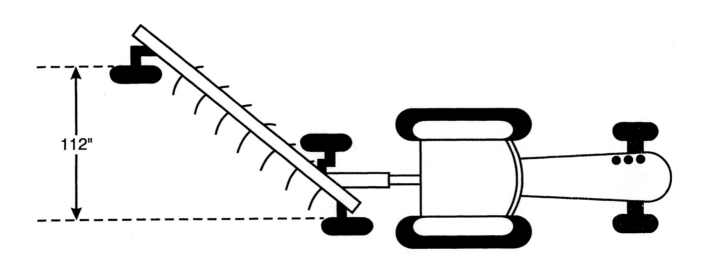

Figure 1–2

_____
14

Exercise 1–2

COMBINED OPERATIONS WITH WHOLE NUMBERS

Review Section G, page 5, for rules governing the sequence of mathematical operations.

Perform the indicated operations:

1. $8 + 16 \div 2 \times 5 - 6 =$ _____.

2. $(8 + 16) \div [(2 \times 5) - 6] =$ _____.

3. $15 - 4 \times 3 + 6 \div 3 =$ _____.

4. $54 \div [(7 - 4) \times 3] =$ _____.

5. $8 [4 + (11 - 7) \times 3] \div 2 =$ _____.

6. 18 inches $\times$ (16 inches + 25 inches) = _____ square inches.

7. 2×29 hectares + 45 hectares = _____ hectares.

8. (27 square yards + 39 square yards) $\div$ 6 yards = _____ yards.

9. (95 square meters + 275 square meters) $\times$ 3 meters = _____ cubic meters.

10. 7 feet + 4(8 feet + 24 feet) = _____ feet.

11. A group of six steer calves weighs a total of 3,156 pounds. If five of the calves weigh 478 pounds, 558 pounds, 525 pounds, 492 pounds, and 543 pounds, what is the weight of the sixth calf?

_____

_____

12. Figure 1–3 shows the dimensions of an irregular area to be enclosed for a sheep pasture. How many 110-meter rolls of 115-centimeter-high woven-wire fence are needed?

_____

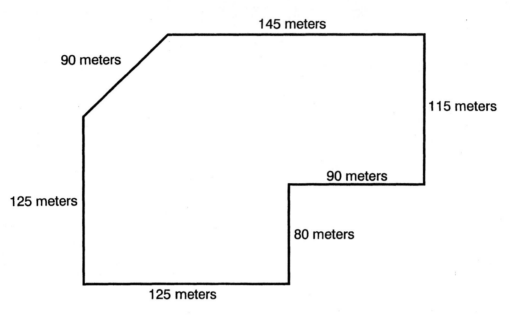

145 meters

90 meters

115 meters

90 meters

125 meters

80 meters

125 meters

Figure 1–3

13. Peanut acreage in the United States is about 600,000 hectares annually. If the average yield is 2,800 kilograms per hectare, what is the annual United State production of peanuts in metric tons? (One metric ton = 1,000 kilograms.)

_____

14. An orchard has 17 rows of pear trees, each row having 29 trees. If each tree is expected to produce 23 bushels of pears this year, what is the potential yield from the orchard?

_____

15. An implement dealer agrees to sell a tractor for the cash price of $45,678. If the dealer gives an $18,400 trade-in allowance for a used tractor and the buyer has $7,450 available for the purchase, how much will need to be borrowed to make the payment?

_____

_____ _____

Name_____   Date_____

16. Figure 1–4 shows a section of farm field map and the accompanying record of the year's crop production.

    a. What was the total yield from the four corn fields?   _____

    b. How many metric tons of corn were harvested?
       (One metric ton = 1,000 kilograms.)   _____

    c. What was the average yield from the four corn fields?
       (Give the yield in kilograms per hectare.)   _____

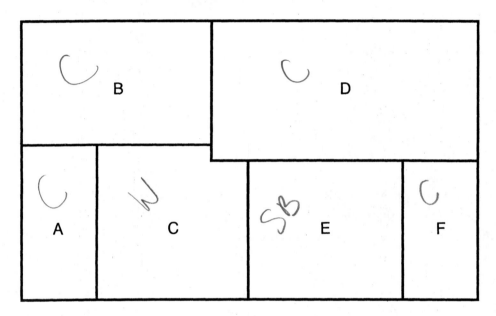

Figure 1–4

Field	Crop	Area (Hectares)	Yield (Kilograms)
A	Corn	18	153,395
B	Corn	35	311,455
C	Wheat	35	106,097
D	Corn	56	382,912
E	Soybeans	32	75,447
F	Corn	16	108,238

17. A truck-garden operator harvested 8,166 pounds of onions. If 230 bags weighing 25 pounds each are packed for a local grocer, what weight of onions remains?

_____

18. Colony cages, having a capacity of 24 hens each, are used in housing a flock of 98,400 laying hens.

   a. How many cages are needed for the flock?   _____

   b. If building and equipment costs are $6.00 per bird, what is the total cost of the facilities to house this flock?   _____

19. Combine A has a harvesting capacity of 620 bushels of corn per hour. Combine B has a 495 bushel-per-hour capacity. How many more bushels of corn can Combine A harvest during nine hours of operation?

_____

20. Figure 1–5 shows plans for a 24-sow farrowing house. Twelve farrowing units, having the dimensions of the unit shown, are to be installed on each side of the alleyway. How long does the building's interior have to be to accommodate these 12 farrowing units?

_____

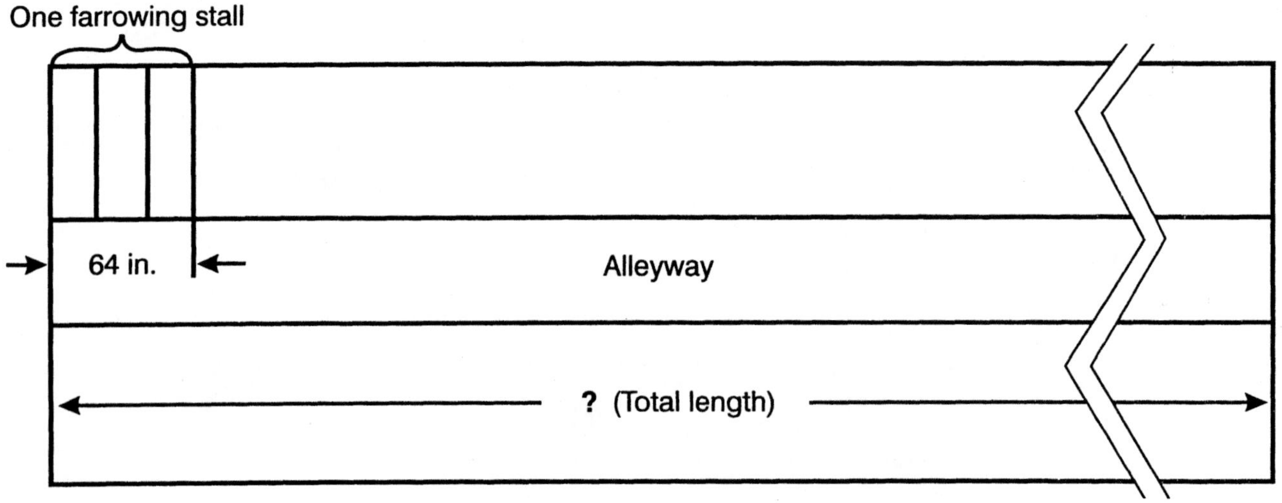

One farrowing stall

64 in. ← Alleyway

? (Total length)

Figure 1–5

21. A litter of nine pigs weighed a total of 141 pounds. After 10 days, the litter weighs 173 pounds. What is the weight gain per pig?

_____

22. A farmer's truck holds 320 bushels. If the farmer fills that truck four times in a 40-acre field of soybeans, what is the yield per acre?

_____

23. Figure 1–6 shows a plan for a new feedlot. Sections AB, BC, and DE of the fenceline are to be fenced with 5-foot woven-wire fence. If a total of 8 feet of fence is allowed for fastening at the end posts, what total length of woven-wire fence is needed?

_____

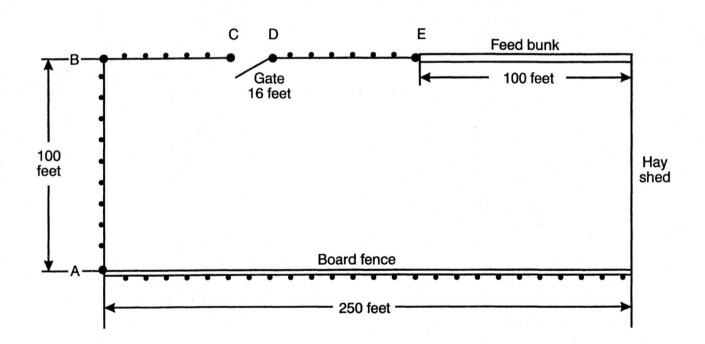

Figure 1–6

24. Alfalfa hay is fed to a herd of 60 dairy cows at the rate of 14 pounds per head per day.

   a. How many bales of hay are needed each day if the hay averages
      70 pounds per bale?   _____

   b. What total number of bales of hay is needed if hay is fed at this
      rate for seven months? (Use a 30-day month.)   _____

25. Last week a grain terminal shipped 25 truckloads of spring wheat, each averaging 22,000 kilograms.

   a. Calculate the total number of kilograms of wheat shipped.   _____

   b. What was the weight of the wheat in metric tons?
      (One metric ton = 1,000 kilograms.)   _____

   c. A river barge can carry 2,815 metric tons of grain. How
      many of these truckloads are needed to fill a barge?   _____

Name_____ Date_____

Exercise 1–3

APPLICATION EXTENSION

To achieve maximum **yield per acre** for many crops, the proper plant population must be calculated.

Plant Population

Example: Seeds are spaced 11 inches apart on 36-inch rows. How many seeds are planted per acre?

Step 1: Calculate the linear feet in 1 acre.

Square feet in 1 acre ÷ row width (in feet) = linear feet per acre

43,560 ÷ 3 = 14,520 linear feet per acre

Step 2: Calculate the linear inches in 1 acre.

Linear feet × 12 inches per foot = linear inches per acre

14,520 × 12 = 174,240 linear inches per acre

Step 3: Calculate number of seeds per acre.

Linear inches ÷ distance of seeds in inches = seeds per acre

174,240 ÷ 11 = 15,840 seeds per acre

1. If seeds are spaced 10 inches apart and row width is 36 inches, how many seeds are planted per acre?

_____

2. If seeds are spaced 9 inches apart and row width is 24 inches, how many seeds are planted per acre?

_____

2

COMMON FRACTIONS

Key Terms

denominator
equivalent fraction
fraction
improper fraction
invert the divisor
least common multiple (LCM)

lowest common denominator (LCD)
lowest terms
mixed number
numerator
prime number
proper fraction

Career Connections

WATER FARMING

Some animals and plants grow on land; others grow in water and are known as aquatic organisms. Many aquatic organisms, such as fish, shrimp, and oysters, are popular as food.

Natural supplies of some fish and other species have declined and the streams, lakes, and oceans cannot meet consumer demand. This has resulted in water farming. Water farming, known as aquaculture, is growing crops in water environments. Most of these species are raised in ponds, but others are kept in tanks and cages. Since aquatic species are quite different from those grown on the land, farmers need new skills to be successful.

Proper water management is vitally important to the aquafarmer. Dissolved oxygen must be maintained at the appropriate level for a species. Ammonia and nitrates are monitored and regulated. Water pH and minerals may need to be controlled. In closed systems, filtration is important to remove suspended solid materials. Other water factors may need to be considered, depending on the species.

Anyone involved in aquaculture must have a good knowledge of mathematics and chemistry. A number of colleges are now offering specialized courses in aquaculture.

A. Definitions

1. A ***fraction*** is a number expressed in the form $\frac{a}{b}$, where a and b are whole numbers and b is not equal to zero. A fraction represents a portion or part of a whole quantity.

2. The **lower number** of the fraction is called the ***denominator*** and represents the number of equal parts into which the whole is divided.

3. The **upper number** is called the ***numerator*** and represents the number of equal parts with which we are working.

4. A ***proper fraction*** is one in which the **numerator is smaller than the denominator.**

   Example: $\frac{1}{6}$, $\frac{2}{3}$, and $\frac{4}{5}$ are proper fractions.

5. An ***improper fraction*** is one in which the **numerator is larger than the denominator;** that is, it represents more than one whole quantity.

   Example: $\frac{12}{5}$, $\frac{16}{11}$, and $\frac{8}{5}$ are improper fractions.

   An improper fraction is sometimes more meaningful when written as a ***mixed number***. A mixed number is a whole number and a fraction. To convert an improper fraction to a mixed number, divide the numerator by the denominator and express the remainder as a fraction.

   Example: $\frac{13}{5}$ means $13 \div 5 = 2$ with a remainder of 3 and is expressed as $2\frac{3}{5}$.

   The fractions in the example above would be written as:

   $$\frac{12}{5} = 2\frac{2}{5}, \quad \frac{16}{11} = 1\frac{5}{11}, \text{ and } \quad \frac{8}{5} = 1\frac{3}{5}$$

   To change a **mixed number** into an **improper fraction,** find the product of the denominator of the fraction and the whole number and add to it the numerator of the fraction. Use this new number as the numerator of the new fraction.

   Example: $2\frac{7}{8} = 8 \times 2 + 7 = 16 + 7$ or $\frac{23}{8}$.

6. An ***equivalent fraction*** results when the numerator and the denominator are both multiplied or exactly divided by the same non-zero number. Multiplication results in an equivalent fraction said to be in larger terms, whereas division results in an equivalent fraction said to be in lower terms. A fractional answer should always be expressed in ***lowest terms***, which means that the lowest common divisor of the numerator and the denominator is one (1), unless otherwise specified.

Example: $\dfrac{16 \times 2}{24 \times 2} = \dfrac{32}{48}$ (equivalent fraction in larger terms).

$\dfrac{16 \div 8}{24 \div 8} = \dfrac{2}{3}$ (equivalent fraction in lower terms and also lowest terms).

B. Addition of Common Fractions

1. **Fractions having like or common denominators:** To add fractions having like denominators, simply add the numerators and use that sum as the numerator of the new fraction.

   Example: $\dfrac{2}{13} + \dfrac{8}{13} = \dfrac{10}{13}$.

2. **Fractions having unlike denominators:** In order to perform addition of fractions having unlike denominators, each fraction must be changed into an equivalent fraction with all fractions to be added having the same denominator. The denominator must be equal to a common multiple of all of the whole numbers that make up the denominators of the fractions to be added. Calculations are usually simplified when we use the **least common multiple (LCM)** for the denominator. We generally call this number the **lowest common denominator (LCD)** when working with the addition and subtraction of fractions.

   The LCD of a set of denominators is usually found by the prime factor method. Which makes use of prime numbers. A **prime number** is a counting number that is only divisible by itself and the number one.

   Example: Find the LCD of denominators 12, 18, and 20 by the prime-factor method.

   Step 1. Find the prime factors in each number.

   $$12 = 2 \times 2 \times 3$$
   $$18 = 2 \times 3 \times 3$$
   $$20 = 2 \times 2 \times 5$$

   Step 2. Use each prime number as a factor the greatest number of times that it appears in any one of the denominators. The prime factor 2 appears twice in 12 and in 20, so we need to use 2 as a factor twice; 3 appears twice in 18, so we need to use it as a factor twice; and 5 appears in 20 once, so we need to use it as a factor once.

   $$LCD = 2 \times 2 \times 3 \times 3 \times 5 = 180.$$

Example: Find the LCD of 24, 16, 20, and 15.

$$24 = 2 \times 2 \times 2 \times 3$$
$$16 = 2 \times 2 \times 2 \times 2$$
$$20 = 2 \times 2 \times 5$$
$$15 = 3 \times 5$$

$$LCD = 2 \times 2 \times 2 \times 2 \times 3 \times 5$$

Addition of fractions having unlike denominators is accomplished by changing each of the fractions into equivalent fractions having the same denominator by using the LCD. The numerator of the new fraction is found by adding the numerators of the equivalent fractions.

Example: Add $\frac{3}{8}$, $\frac{5}{12}$, and $\frac{2}{15}$.

Find the LCD of 8, 12, and 15.

$$8 = 2 \times 2 \times 2; \ 12 = 2 \times 2 \times 3; \ 15 = 3 \times 5$$
$$LCD = 2 \times 2 \times 2 \times 3 \times 5 = 120$$

$$\frac{3 \times 15}{8 \times 15} = \frac{45}{120}; \ \frac{5 \times 10}{12 \times 10} = \frac{5}{120}; \ \frac{2 \times 8}{15 \times 8} = \frac{16}{120}$$

$$45 + 50 + 16 = 111 \text{ or } \frac{111}{120}.$$

Example: Add $9\frac{3}{8}$, $21\frac{5}{6}$, and $19\frac{2}{9}$.

It is often more convenient to arrange mixed form addends in a column, making equivalent fractions only of the fractional part.

Using prime-factor method of finding LCD:

$$8 = 2 \times 2 \times 2; \ 6 = 2 \times 3; \ 9 = 3 \times 3$$
$$2 \times 2 \times 2 \times 3 \times 3 = 72$$

$$9\frac{3}{8} = 9\frac{27}{72}$$

$$21\frac{5}{6} = 21\frac{60}{72}$$

$$19\frac{2}{9} = 19\frac{16}{72}$$

$$49\frac{103}{72} \text{ or } 50\frac{31}{72}.$$

C. Subtraction of Common Fractions

1. ***Fractions having like or common denominators:*** To find the difference of two fractions having like denominators, simply subtract the numerator of the subtrahend from the numerator of the minuend and use the difference as the numerator of the new fraction.

   Example: $\dfrac{11}{17} - \dfrac{8}{17} = \dfrac{3}{17}$.

2. ***Fractions having unlike denominators:*** To find the difference of fractions having unlike denominators, change each fraction into an equivalent fraction with both fractions having the same LCD. Find the difference of the numerators of the equivalent fractions and use the difference as the numerator of the new fraction. If, when working with fractions in mixed form, the fraction of the minuend is smaller than the fraction of the subtrahend, borrow "one" from the whole number part of the minuend and add it as an equivalent fraction to the fractional part of the minuend. Now the difference of the numerators can be found and used as the numerator of the new fraction.

   Example: Subtract $\dfrac{8}{18}$ from $\dfrac{7}{12}$.

   LCD: $12 = 2 \times 2 \times 3$; $18 = 2 \times 3 \times 3$
   LCD $= 2 \times 2 \times 3 \times 3 = 36$

   $\dfrac{7}{12} \times \dfrac{3}{3} - \dfrac{8}{18} \times \dfrac{2}{2} = \dfrac{21}{36} - \dfrac{16}{36} = \dfrac{5}{36}$.

   Example: Subtract $722\dfrac{2}{3}$ from $979\dfrac{2}{7}$.

   LCD $= 3 \times 7 = 21$

   $$979\dfrac{2}{7} = \quad 979\dfrac{6}{21} = 978\dfrac{21}{21} + \dfrac{6}{21} = 978\dfrac{27}{21}$$
   $$-\,722\dfrac{2}{3} = -\,722\dfrac{14}{21} = \qquad\qquad\quad \dfrac{-\,722\dfrac{14}{21}}{256\dfrac{13}{21}}.$$

D. Multiplication of Common Fractions

To find the product of numbers expressed as fractions, multiply the numerators of the fractions for the numerator of the product and multiply the denominators for the denominator of the product. Express the product in lowest terms.

Example: $\dfrac{2}{3} \times \dfrac{4}{7} = \dfrac{2 \times 4}{3 \times 7} = \dfrac{8}{21}.$

If the computation involves fractions in mixed form, convert the fractions to improper fractions before multiplying.

Example: $3\dfrac{3}{8} \times 1\dfrac{7}{9} = \dfrac{27}{8} \times \dfrac{16}{9} = \dfrac{432}{72} = 6.$

Products of fractions can sometimes be found more quickly and with simpler computations if a process called "cancellation" is used prior to multiplication. It effectively takes the place of the "reducing to lowest terms" step.

First, convert all mixed numbers to improper fractions and then divide all common factors from the numerators and denominators.

Example: Multiply $\dfrac{1}{7} \times 2\dfrac{1}{3} \times 3\dfrac{3}{8} \times 12.$

$$\dfrac{1}{\cancelto{1}{7}} \times \dfrac{\cancelto{1}{7}}{\cancelto{1}{3}} \times \dfrac{\cancelto{9}{27}}{\cancelto{2}{8}} \times \dfrac{\cancelto{3}{12}}{1} = \dfrac{27}{2} = 13\dfrac{1}{2}.$$

When finding the product of mixed-form fractions having large whole numbers, a five-step procedure can be used.

Example: Multiply $244\dfrac{2}{3}$ by $369\dfrac{3}{4}.$

Step 1. Multiply the whole-number parts together.

$244 \times 369 = 90{,}036$

Step 2. Multiply the whole-number part of the multiplier by the fractional part of the multiplicand.

$$\overset{123}{\cancel{369}} \times \dfrac{2}{\cancel{3}} = 123 \times 2 = 246$$

Step 3. Multiply the whole-number part of the multiplicand by the fractional part of the multiplier.

$$\overset{61}{\cancel{244}} \times \dfrac{3}{\cancel{4}} = 61 \times 3 = 183$$

Step 4. Multiply the fractional parts together.

$$\frac{\overset{1}{\cancel{2}}}{\underset{1}{\cancel{3}}} \times \frac{\overset{1}{\cancel{3}}}{\underset{2}{\cancel{4}}} = \frac{1}{2}$$

Step 5. Add the four products together to get the total product.

$$
\begin{array}{r}
90,036 \\
246 \\
183 \\
+ \quad \dfrac{1}{2} \\
\hline
90,465\dfrac{1}{2}.
\end{array}
$$

E. Division of Common Fractions

In the discussion of equivalent fractions, it was stated that multiplying both the numerator and the denominator of a fraction by the same non-zero number produces a new equivalent fraction. The following example uses this concept to explain the operation of the division of fractions.

Example: Divide $\frac{2}{3}$ by $\frac{1}{6}$.

$\frac{2}{3}$ is the dividend and $\frac{1}{6}$ is the divisor. The number chosen as the multiplier is the reciprocal of the divisor. To find the reciprocal of a fraction, invert the fraction (interchange the numerator and the denominator). The product of a number and its reciprocal is one (1).

$$\frac{\dfrac{2}{3} \times \dfrac{6}{1}}{\dfrac{1}{6} \times \dfrac{6}{1}} = \frac{\dfrac{12}{3}}{\dfrac{6}{6}} = \frac{4}{1} = 4.$$

The foregoing presentation shows the basis for the development of the inverted-divisor method of dividing fractions. To find the quotient of fractional numbers by the inverted-divisor method, **invert the divisor** (find the reciprocal of the divisor) **and multiply**. The process of cancellation is useful in the simplification of computations in the division of fractions.

Example: Divide $\frac{1}{4}$ by $\frac{3}{5}$.

$$\frac{1}{4} \times \frac{5}{3} = \frac{5}{12}.$$

Example: Divide $\frac{3}{4}$ by 6.

$$\frac{\overset{1}{\cancel{3}}}{4} \times \frac{1}{\underset{2}{\cancel{6}}} = \frac{1}{8}.$$

Example: Divide $12\frac{8}{9}$ by $4\frac{5}{6}$.

$$\frac{116}{9} \div \frac{29}{6}; \quad \frac{\overset{4}{\cancel{116}}}{\underset{3}{\cancel{9}}} \times \frac{\overset{2}{\cancel{6}}}{\underset{1}{\cancel{29}}} = \frac{4 \times 2}{3 \times 1} = \frac{8}{3} = 2\frac{2}{3}.$$

F. Combined Operations with Fractions

In simplifying problems involving several operations with fractions, signs of grouping, such as parenthesis, division bars, etc., indicate which mathematical procedures to perform first. The final result should be a whole number or a single fraction.

Example: Simplify $\dfrac{11\frac{1}{3}}{9\frac{1}{3} - (2\frac{1}{2} \times 1\frac{7}{15})}$.

$$\frac{\dfrac{34}{3}}{\dfrac{28}{3} - (\dfrac{\overset{}{\cancel{5}}}{\underset{1}{\cancel{2}}} \times \dfrac{\overset{11}{\cancel{22}}}{\underset{3}{\cancel{15}}})} = \frac{\dfrac{34}{3}}{\dfrac{28}{3} - \dfrac{11}{3}} = \frac{\dfrac{34}{3}}{\dfrac{17}{3}} = \frac{\overset{2}{\cancel{34}}}{\cancel{3}} \times \frac{\overset{1}{\cancel{3}}}{\underset{1}{\cancel{17}}} = 2.$$

Exercise 2–1

ADDITION AND SUBTRACTION OF COMMON FRACTIONS

Write equivalent fractions in lowest terms for the following:

1. $\dfrac{27}{36} =$

2. $\dfrac{140}{420} =$

3. $\dfrac{350}{2,100} =$

4. $\dfrac{56}{98} =$

5. $\dfrac{186}{558} =$

Write equivalent fractions with denominators or numerators as indicated.

6. $\dfrac{8}{26} = \dfrac{?}{39}$ $? =$ _____.

7. $\dfrac{3}{5} = \dfrac{21}{?}$ $? =$ _____.

8. $\dfrac{11}{12} = \dfrac{?}{84}$ $? =$ _____.

Write as equivalent numerals in mixed number form:

9. $\dfrac{95}{9} =$

10. $\dfrac{309}{15} =$

11. $\dfrac{201}{19} =$

Write as improper fractions in lowest terms:

12. $11\frac{9}{10} =$

13. $361\frac{2}{3} =$

14. $45\frac{7}{8} =$

Find the LCD for the following fractions:

15. $\frac{7}{24} + \frac{3}{15} + \frac{5}{16} + \frac{7}{8}$ LCD = _____.

16. $\frac{6}{54} + \frac{2}{27} + \frac{1}{9} + \frac{3}{16}$ LCD = _____.

17. $\frac{7}{18} + \frac{9}{25} + \frac{11}{24} + \frac{5}{27}$ LCD = _____.

Perform the indicated operations, expressing the result in lowest terms:

18. $6\frac{3}{8}$ pounds $+ 8\frac{1}{2}$ pounds = _____.

19. $7\frac{1}{16}$ feet $+ 8\frac{1}{4}$ feet = _____.

20. $30\frac{1}{4}$ square yards $+ 25\frac{1}{3}$ square yards = _____.

21. $19\frac{1}{6}$ quarts $- 12\frac{3}{8}$ quarts = _____.

22. $15\frac{2}{3}$ fluid ounces $- 4\frac{2}{5}$ fluid ounces = _____.

23. $26\frac{1}{4}$ square miles $- 16\frac{1}{3}$ square miles = _____.

24. Which is smaller, $\frac{2}{3}$ acre or $\frac{11}{16}$ acre?   _____

32

Name_____ Date _____

25. Arrange in order of size from largest to smallest: $\frac{37}{16}$ inches, $\frac{19}{8}$ inches, and $\frac{9}{4}$ inches.

   _____

26. Which quantity is the largest, $\frac{3}{4}$ inch, $\frac{7}{8}$ inch, or $\frac{2}{3}$ inch?   _____

27. Arrange in order of size from largest to smallest: $\frac{3}{5}$ quart, $\frac{2}{3}$ quart, and $\frac{5}{8}$ quart.

   _____

Add the following mixed numbers:

28. $\quad 4\frac{2}{3}$

$\quad\quad 93\frac{1}{8}$

$\quad\quad 27\frac{5}{12}$

$\quad + \ \ 6\frac{6}{16}$

29. $\quad 21\frac{11}{15}$

$\quad\quad 63\frac{5}{12}$

$\quad\quad 4\frac{7}{30}$

$\quad + \ \ 19\frac{3}{4}$

30. $\quad 14\frac{6}{24}$

$\quad\quad 105\frac{11}{20}$

$\quad\quad 28\ \frac{3}{8}$

$\quad + \ \ 77\frac{7}{15}$

Subtract the following mixed numbers:

31. $104\frac{7}{15}$

$\quad - \ \ 17\ \frac{7}{9}$

32. $\quad 87\frac{3}{5}$

$\quad - \ \ 53\frac{5}{12}$

33. $\quad 921\frac{5}{11}$

$\quad - \ \ 109\frac{2}{3}$

Solve the following problems:

34. Plowing 8 inches deep on gumbo soil requires a tractor with $13\frac{1}{3}$ drawbar horsepower per foot of width plowed. Doing the same operation on sandy soil requires 4 drawbar horsepower per foot of width. How much less are the power requirements on sandy soil?

   _____

_____

33

35. If a market lamb on full feed eats an average of $1\frac{2}{3}$ pounds of corn silage, $1\frac{1}{8}$ pounds of legume hay, and $1\frac{1}{4}$ pounds of grain concentrate per day, what is the total weight of daily feed intake?

_____

36. The hydraulic pump hose on a tractor has an outside diameter of $1\frac{1}{8}$ inches. If the wall thickness of the hose is $\frac{11}{32}$ inch ($\frac{22}{32}$ inch for both walls), what is the inside diameter of the hose?

_____

37. A beehive has exterior dimensions as given in Figure 2–1. If the walls are constructed of $\frac{7}{8}$ inch timber, what is the interior length and width of the hive?

_____

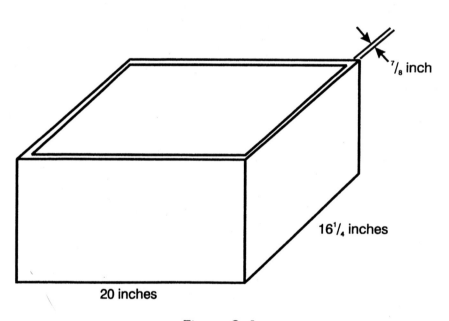

$\frac{7}{8}$ inch

$16\frac{1}{4}$ inches

20 inches

Figure 2–1

38. A fence is to be constructed using cedar fence posts. The corner post has a 7-inch diameter and is set $3\frac{1}{4}$ feet into the ground. If $5\frac{7}{12}$ feet of post are above ground, what is the total length of the post?

_____

_____

34

Name____________________ Date________________

39. If a $6\frac{1}{8}$-pound roaster chicken yields a $4\frac{2}{5}$-pound ready-to-cook carcass, what is the loss of weight from dressing and drawing?

____

40. A student picks apples for an apple grower after school hours. If the student works $2\frac{1}{3}$ hours, $1\frac{3}{4}$ hours, $2\frac{1}{2}$ hours, $2\frac{2}{3}$ hours, $3\frac{1}{6}$ hours, and $9\frac{1}{4}$ hours during a week, how many hours did the student work?

____

41. A board fence is to be built around a paddock. Figure 2–2 shows an 8-foot section of the fence with the dimension lumber and spacings to be used. What is the overall height of the fence?

____

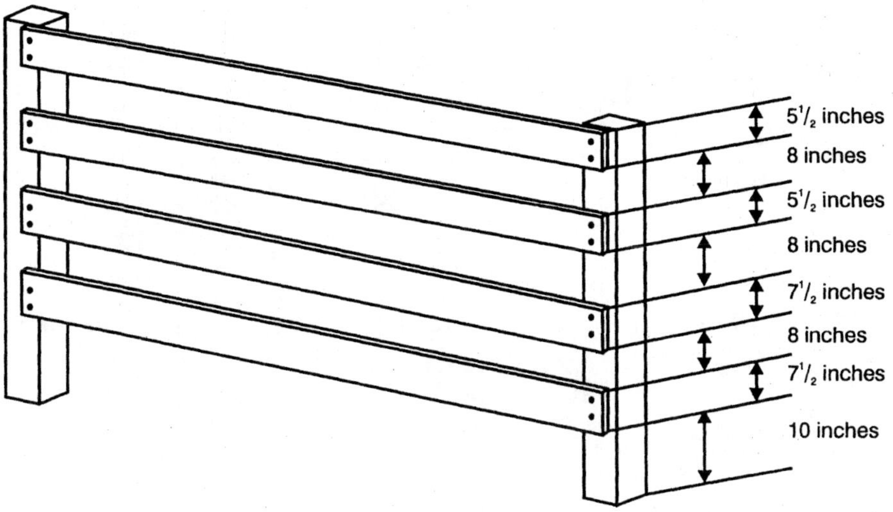

$5\frac{1}{2}$ inches
8 inches
$5\frac{1}{2}$ inches
8 inches
$7\frac{1}{2}$ inches
8 inches
$7\frac{1}{2}$ inches
10 inches

Figure 2–2

42. Partner A of a farm partnership owns $\frac{3}{7}$ of the dairy herd while partner B owns $\frac{4}{9}$ of the herd. Which partner owns the larger share and how much larger is the share?

____

____

43. A citrus grower produced 115¾ tons of fruit from one grove and 167⅔ tons from another. How many tons of fruit were produced from the two groves?

_____

44. What is the shortest length machine bolt that can be used to fasten together the two parts of the plow frame shown in Figure 2–3?

_____

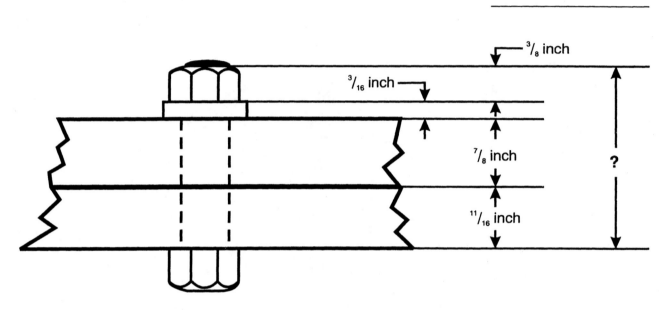

Figure 2–3

45. Acreage of soybeans had an average yield of 46⅔ bushels per acre after harvest losses. Estimated losses are 2⅘ bushels per acre. What was the actual yield per acre before losses?

_____

46. If broiler chicks cost 25⅞ cents per bird in March and cost 32¼ cents each in May, how much less does a chick cost when purchased in March?

_____

47. An automatic hog waterer is to be installed. Three lengths of pipe are needed: 10½ inches, 22⅔ inches, and 15¾ inches. What total length of pipe is needed if $\frac{3}{16}$ inch (total) of pipe is wasted by cutting the pipe?

_____

48. The length of machine part consists of four pieces measuring 8 $\frac{3}{16}$ inches, 5¼ inches, 15⅞ inches, and 1$\frac{15}{32}$ inches. Find the total length of the machine part.

_____

49. Value of a beef carcass is partially dependent upon size of the rib eye. Figure 2–4 shows the rib eye of two different grades of beef carcasses. What is the difference in rib eye area of the two grades of carcasses?

_____

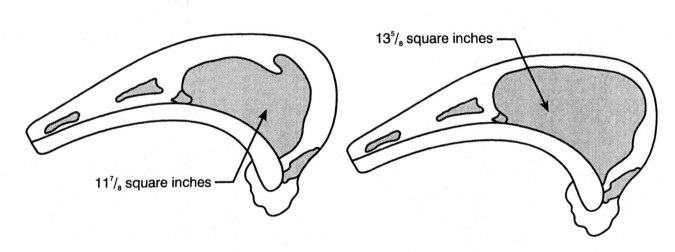

13⅝ square inches

11⅞ square inches

Figure 2–4

50. In one area of the United States average mohair clip from Angora goats is 6⅗ pounds. In another area, the mohair clip is 3⅔ pounds per goat. Find the difference in the weight of mohair clip per goat in the two production areas.

_____

Exercise 2–2

MULTIPLICATION AND DIVISION OF COMMON FRACTIONS

1. $\dfrac{5}{6} \times \dfrac{30}{45} =$

2. $\dfrac{4}{9} \times \dfrac{3}{14} \times \dfrac{7}{16} =$

3. $\dfrac{11}{14} \times \dfrac{1}{9} \times \dfrac{28}{33} =$

4. $2 \times 5\dfrac{1}{6} \times 0 =$

5. $2\dfrac{2}{3} \times \dfrac{5}{6} \times 4\dfrac{1}{8} =$

6. $\dfrac{45}{6} \times 6\dfrac{3}{5} \div 11 =$

7. $3\dfrac{6}{13} \times 7\dfrac{4}{5} \div 5\dfrac{2}{5} =$

8. $6\dfrac{3}{4} \div 1\dfrac{1}{8} =$

9. $57 \div 3\dfrac{4}{5} =$

10. $3\dfrac{3}{19} \div 3\dfrac{3}{4} =$

11. $3\dfrac{3}{7} \times 1\dfrac{3}{4} \div 3\dfrac{3}{5} =$

12. $\dfrac{3}{4} \div 8 =$

13. $2\dfrac{1}{13} \times \dfrac{3}{4} \times 4\dfrac{1}{3} =$

14. $\dfrac{3}{8} \times \dfrac{4}{9} \times \dfrac{4}{11} =$

15. $\dfrac{2}{9} \div 4\dfrac{3}{8} \times 2\dfrac{1}{3} \div 5\dfrac{1}{3} =$

16. $1\dfrac{3}{4} \div 6\dfrac{3}{4} =$

17. $\dfrac{5}{6} \times 12 \times \dfrac{1}{4} =$

18. $4\dfrac{3}{5} \times \dfrac{5}{7} \div 5\dfrac{1}{3} \times 5\dfrac{3}{5} =$

Multiply using the five-step method.

19. $321\frac{5}{6}$
    $162\frac{2}{3}$

20. $117\frac{5}{8}$
    $104\frac{4}{9}$

Perform the indicated operations, expressing results in lowest terms.

21. $30\frac{1}{4}$ rods $\times$ $2\frac{2}{11}$ rods = _____ square rods.

22. 9 inches is what fractional part of 1 foot? _____

23. $1\frac{3}{5}$ of $18\frac{3}{4}$ quarts = _____ quarts.

24. $22\frac{5}{16}$ pounds = _____ ounces.

25. How many pecks are in $16\frac{1}{2}$ bushels? _____

26. $13\frac{1}{3}$ cubic yards $\div$ $\frac{8}{9}$ yard = _____ square yards.

27. 16 acres is $\frac{2}{3}$ of what number of acres? _____

28. $147\frac{3}{8}$ miles $\div$ $4\frac{1}{2}$ gallons = _____ miles per gallon.

29. How many $\frac{1}{8}$ inches are in $6\frac{3}{4}$ inches? _____

30. $9\frac{3}{4}$ square feet $\div$ $4\frac{1}{3}$ feet = _____ feet.

Solve the following problems:

31. Oats make up ⅖ of the weight of mixed horse feed. A truckload of this mixture contains 1,600 pounds of oats. How many pounds of mixed feed are in the truckload?

_____

32. If 7½ gallons of fuel require 1 cubic foot of space, what is the volume in cubic feet of a fuel storage tank designed to hold 1,000 gallons of fuel?

_____

33. An insecticide concentrate contains 7 pounds of actual chemical per gallon. If ⅛ of a pound of actual chemical is to be applied per acre, how many acres can be treated with 1 gallon of the insecticide?

_____

34. Figure 2–5 shows a length of paddock fence to be built using cedar posts placed 8¼ feet apart. How many posts are needed for the fence?

_____

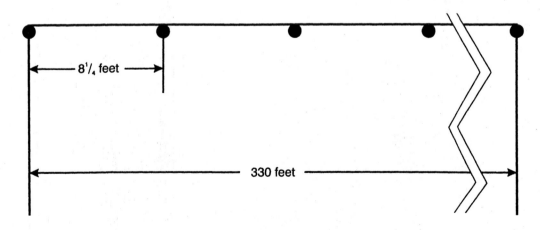

8¹⁄₄ feet

330 feet

Figure 2–5

35. A tractor pulling a 14-foot disk at 5 miles per hour can till 8⅘ acres per hour. How many acres can be disked in 6¾ hours?

_____

36. A hog feeding floor is 54⅔ feet long and 27¾ feet wide. Find the area of the floor by multiplying the length by the width.

_____

37. Grain swaths averaging 13⅔ feet in width can be cut by a skilled operator using a 14-foot self-propelled windrower. At this rate, how many swaths would be cut on a field 1,353 feet wide?

_____

38. A stable manager purchased a roll of ½-inch cotton rope for making lead ropes. How many lead ropes (lengths as indicated in Figure 2–6) can be cut from the roll of rope?

_____

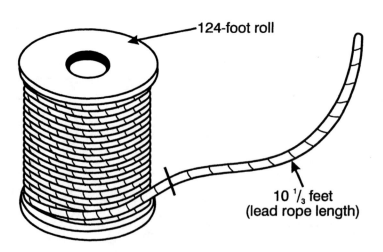

124-foot roll

10 ⅓ feet
(lead rope length)

Figure 2–6

39. An average of 76½ pounds of oats are planted per acre. If oats weigh 34 pounds per bushel, how many bushels of oats are planted per acre?

_____

40. A forage harvester consumes 1⅓ gallons of fuel per acre and harvests 3½ acres per hour. What amount of fuel is consumed per hour?

_____

41. Receipts show that a turkey grower is receiving an average of 45 cents per pound of live weight. This price is ¼ more than was received last year. What was last year's price?

_____

42. A variety of peanut crushed for oil has the content shown in Figure 2–7. How many pounds of oil could be expected from a ton of clean, mature peanuts?

_____

COMPOSITION OF PEANUTS

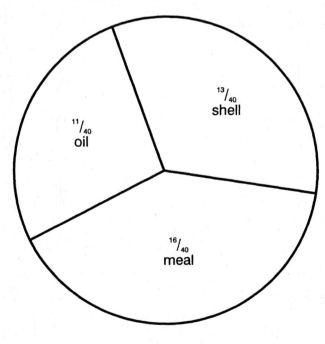

Figure 2–7

43. In the process of dehydration of potatoes for instant mashed potatoes, approximately $^{19}/_{25}$ of the weight of the potato is removed. How many pounds of dehydrated potatoes result from the processing of a bushel of potatoes? (One bushel of potatoes weighs 60 pounds.)

_____

44. Approximately $^{1}/_{6}$ of the liveweight of a beef steer becomes hamburger. At this rate, how much hamburger would be made from a 1,152-pound steer?

_____

45. Broccoli is processed and packaged for freezing in bags containing 1¼ pounds of broccoli. If 342 bags are ready for freezing, how many pounds of broccoli have been processed?

_____

46. The field capacity of a tractor and plow is 3⅗ acres per hour. Plowing a short irregular field cuts field capacity to ⅚ of its usual capacity. How many acres can be plowed per hour on the short irregular field?

_____

47. A landscape designer recommends that a short decorative fence be put around a circular landscaped area. If the area has the dimensions given in Figure 2–8, how many feet of fence are needed? (Circumference can be found by multiplying the length of the diameter by the constant π. Use $^{22}/_{7}$ for π.)

_____

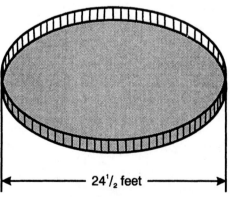

24¹/₂ feet

Figure 2–8

48. From past records, a breeder of dairy goats expects that $^{22}/_{43}$ of the kids born will be male. If 129 kids are born, how many could be expected to be male?

_____

49. A baler making large round bales is used to harvest alfalfa-brome hay. If the average weight of the bales is ¾ ton, how many bales were made when 105 tons of hay were harvested?

_____

50. A tractor and disk can cover 8¼ acres per hour. How many hours will it take to disk 92⅖ acres?

_____

Exercise 2–3

COMBINED OPERATIONS WITH COMMON FRACTIONS

Perform the indicated operations, expressing the resulting fractions in lowest terms. For problems 1 through 15, review Section G on page 5 for rules governing sequence of mathematical operations.

1. $8\frac{1}{3} + 2\frac{1}{2} - 3\frac{1}{4} =$ _____.

2. $\frac{3}{4} \times 3\frac{1}{5} \div 6\frac{3}{10} =$ _____.

3. $10\frac{1}{4} - 1\frac{1}{2} \times \frac{1}{4} + 1\frac{5}{6} =$ _____.

4. $6\frac{1}{8} \div \left[\left(2\frac{1}{3} - 1\frac{5}{6}\right) \times 1\frac{3}{4}\right] =$ _____.

5. $\frac{1}{2}\left[\frac{1}{8} + \left(\frac{1}{4} - \frac{1}{8}\right) \times 1\frac{1}{2}\right] \div 2\frac{1}{2} =$ _____.

6. $2\frac{1}{2}$ inches $\left(3\frac{1}{3}$ inches $+ 5\frac{1}{5}$ inches$\right) =$ _____ square inches.

7. $2\frac{1}{3} \times \frac{2}{5}$ yard $+ 7\frac{1}{3}$ yards $=$ _____ yards.

8. $\left(3\frac{1}{6}$ square feet $+ 16\frac{1}{2}$ square feet$\right) \div \frac{1}{6}$ foot $=$ _____ feet.

9. $\left(4\frac{1}{8}$ square inches $+ 6\frac{3}{4}$ square inches$\right) \times 2\frac{2}{3}$ inches $=$ _____ cubic inches.

10. $5\frac{1}{3}$ rods $+ \frac{1}{3}\left(2\frac{1}{4}$ rods $+ 6\frac{3}{8}$ rods$\right) =$ _____ rods.

Simplify:

11. $\dfrac{\dfrac{5}{6} + \dfrac{4}{9}}{11\dfrac{1}{2} \times 1\dfrac{1}{2}}$

12. $\dfrac{17\dfrac{1}{3}}{5 - \dfrac{1}{2 - \dfrac{1}{2}}}$

13. $\dfrac{(27 \div 5) \times 40}{1\dfrac{1}{5}}$

14. $\dfrac{17\dfrac{5}{8}}{1 + \left(\dfrac{3}{4} \times \dfrac{7}{30}\right)}$

15. $\dfrac{5\dfrac{13}{15}}{1 - \left(\dfrac{4}{9} \times \dfrac{45}{64}\right)}$

Solve the following problems:

16. Fahrenheit temperature can be found by multiplying the Celsius temperature by ⁹⁄₅ and adding 32. (F = ⁹⁄₅ C + 32.) If 80 degrees Celsius produces maximum efficiency of a certain tractor engine, what is the corresponding Fahrenheit temperature for best operation?

Name_____ Date_____

17. A 4-H calf weighed 673⅞ pounds. What will the calf weigh after gaining 2½ pounds per day for 15 days?

_____

18. Real estate taxes on farm land in a certain state are 7⁄500 of the actual land value. What amount of tax is payable on a 600-acre farm valued at $1,200 per acre?

_____

19. When full, a tractor's fuel tank holds 56 gallons of fuel. Figure 2–9 shows the fuel gauge on the tractor. If the gauge is correct, how many gallons of fuel are in the tank?

_____

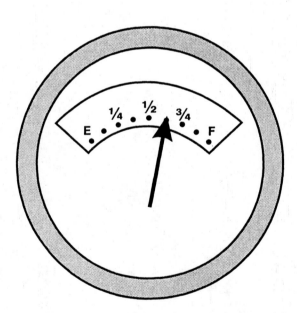

Figure 2–9

20. One picker sheller can harvest 2⅓ acres of corn per hour and another can harvest 1½ acres per hour. How long would it take the two machines working together to harvest 46 acres of corn?

_____

49

21. Maleic hydrazide (MH) is a chemical sprout inhibitor that is mixed with water and sprayed on a potato crop. If 6½ pints of MH are applied per acre, how many gallons of the chemical are needed for 68 acres of potatoes? (8 pints = 1 gallon)

_____

22. Annual inventory of a farmer's total assets sets the value at $850,800. The farmer estimates ⅜ of the value is in land, ⅓ is in buildings, ⅙ is in machinery, and the remainder is in livestock. What is the value of the livestock?

_____

23. Figure 2–10 shows the crushing rollers of a hay conditioner. Find the distance between the centers of the rollers.

_____

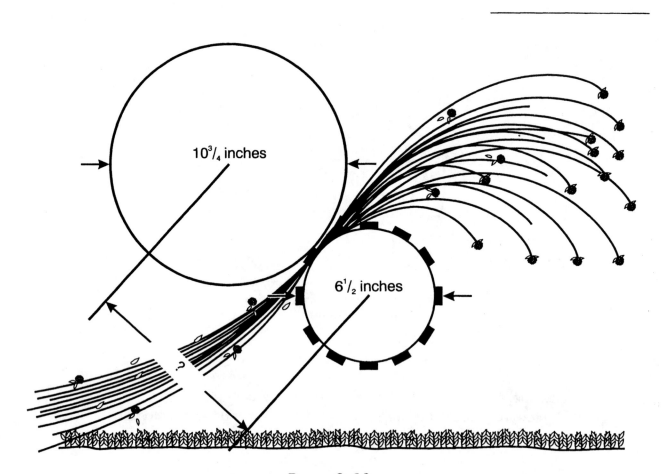

$10^3/_4$ inches

$6^1/_2$ inches

Figure 2–10

24. A 70¼-pound pig is put on a special ration for a 12-day test period. If the pig weighs 83 7⁄12 pounds at the end of the test period, what was the daily rate of gain?

_____

25. The layout of a 600-acre farm purchased by a beef farmer is shown in Figure 2–11. What fraction of the farm is wooded pastureland?

_____

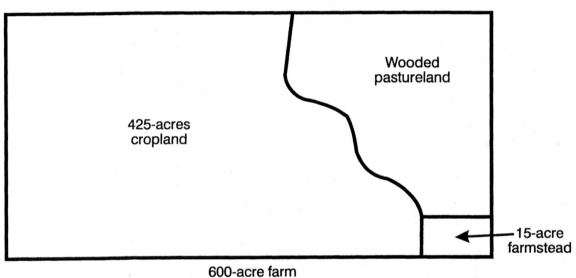

600-acre farm

Figure 2–11

26. If an implement dealer's cost is ⅓ less than the selling price, what is the selling price of a self-propelled cotton picker that costs the dealer $28,000?

_____

27. A total of 2,678 acres of cropland are worked by a family partnership. If 158½ acres are in alfalfa, 68¾ are in spring wheat, 694⅔ are in soybeans, and the remaining acres are in corn, what is the corn acreage?

_____

28. A board 230½ inches long is to be cut into nine pieces of equal length, to be used as shelves for a farm workshop. If each saw cut wastes ⅛ inch, how long is each piece?

_____

29. Figure 2–12 shows a section of auger in a feed conveyor. It is installed in a 120-foot feed bunk. With each revolution of the auger, feed is moved the distance of the flight spacing. If the auger is rotating 120 revolutions per minute, how many minutes will it take haylage to reach the far end of the feed bunk?

_____

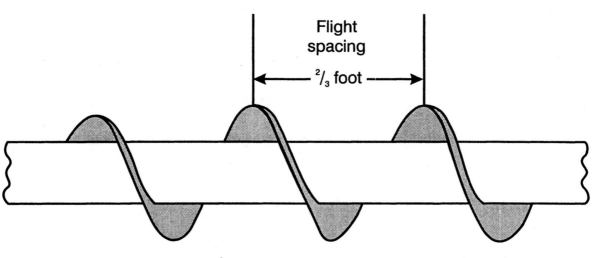

Flight
spacing
←— ⅔ foot —→

Figure 2–12

30. A farm partnership wants to include an additional member in the partnership. Partner A owns ¾ of the total assets. If ⅓ of partner A's assets are sold for $52,000 to the new partner, what is the value of the total assets of the partnership?

_____

3

DECIMALS

Key Terms

addition of decimals
cost estimate
The Decimal System
division of decimals

equivalent decimals and fractions
multiplication of decimals
rounding numbers
subtraction of decimals

Career Connections

CHICKEN TECHNOLOGY

High technology is used in raising chickens! Both egg and meat producers use recent inventions to improve quality and speed production. Researchers study the chicken's needs and how to provide a good environment. Differences in chickens for egg and meat production are carefully noted with new methods being developed that consider the well-being of poultry through improved diet and better health. This helps eliminate loss of birds and provides consumers with top quality poultry products at reasonable prices.

Today, people around the world enjoy chicken because new technology has made large-scale production possible.

A. The Decimal System

The Decimal System is another form of expressing numbers that are parts of whole numbers. The expanded place-value is as follows:

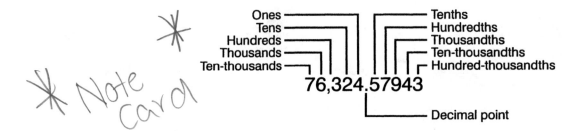

The whole part on the left is separated from the partial part on the right by a decimal point. To distinguish more easily between the various quantities of a decimal number, a name is given to each group of three digits to the left of the decimal point. These groups are usually separated by a comma and given names such as thousands, millions, and billions. The word "and" is used in numeration of the decimal system to indicate the separating point (decimal point) between the whole part of the number and the decimal fraction part. A hyphen (-) is used in writing compound numbers less than one hundred. The place values to the right of the decimal point that are multinamed have a hyphen between the parts of the name.

Examples: 34 is written thirty-four;
0.0062 is written sixty-two ten-thousandths;
1,278.003 is written one thousand, two hundred seventy-eight and three thousandths.

B. Rules for Rounding Numbers

1. Rounding decimals

Step 1.   Determine the digit or the number of decimal places to be retained.

Step 2.   Consider the digit immediately to the right of the digit to be retained.

Step 3.   If the next digit to the right of the digit to be retained is 5 or more, add one to the retained digit and delete all digits to its right. If the next digit is 4 or less, the retained digit is left unchanged and all digits to its right are deleted.

Example: Round 26.483 to hundredths place.

26.483 ≈ 26.48.

Example: Round 85.6256 to thousandths place.

85.6256 ≈ 85.626

2. Rounding whole numbers

Step 1.   Determine the digit to be retained.

Step 2.   Examine the digit immediately to the right of the digit to be retained.

Step 3.   If the next digit is 5 or more, add one to the retained digit and replace the remaining digits to its right with zeros. If the next digit is 4 or less, the retained digit is left unchanged and the remaining digits to its right are replaced with zeros.

Example: Round 34,635 to hundreds place.

34,635 ≈ 34,600.

Example: Round 29,946 to tens place.

29,946 ≈ 29,950.

C. Equivalent Decimals and Fractions

1. To change a fraction to a decimal, divide the numerator by the denominator.

Example: Change $\frac{3}{4}$ to a decimal.

$$
\begin{array}{r}
0.75 \\
4\overline{)3.00} \\
\underline{2\,8} \\
20 \\
\underline{20} \\
00
\end{array}
\qquad
\frac{3}{4} = 0.75.
$$

Example: Change $\frac{2}{3}$ to a decimal (round to hundredths place).

$$
\begin{array}{r}
.666 \\
3\overline{)2.000} \\
\underline{1\,8} \\
20 \\
\underline{18} \\
20 \\
\underline{18} \\
2
\end{array}
\qquad
\frac{2}{3} = 0.666 \text{ or } 0.67.
$$

Example: Change $3\frac{3}{8}$ to a decimal.

Express the mixed number as an improper fraction, then divide the numerator by the denominator.

$$3\frac{3}{8} = \frac{27}{8}$$

$$\begin{array}{r} 3.375 \\ 8\overline{)27.000} \\ \underline{24} \\ 3\,0 \\ \underline{2\,4} \\ 60 \\ \underline{56} \\ 40 \\ \underline{40} \\ 00 \end{array}$$

$$3\frac{3}{8} = 3.375.$$

2. To change a decimal to a fraction:

Step 1.   Count the number of decimal places in the number.

Step 2.   Write the fraction using the number without the decimal point as the numerator. The denominator of the fraction will be a one (1) followed by the number of zeros counted in Step 1.

Example: Change the decimal 0.25 to a fraction.

Step 1.   There are two decimal places.

Step 2.   The numerator of the fraction is 25 and the denominator is a 1 with two zeros.

Step 3. $\dfrac{25}{100} = \dfrac{1}{4}$.

Example: Change 0.625 to a fraction.

Step 1.   There are three decimal places.

Step 2.   The numerator of the fraction is 625 and the denominator is a 1 with three zeros.

Step 3. $\dfrac{625}{1,000} = \dfrac{5}{8}$.

3. To change a decimal that includes a fractional part to a fraction:

Step 1. Change the fractional part to a decimal.

Step 2. Include this answer (without any additional decimal point) at the end of the decimal value.

Step 3. Continue as in the previous examples.

Example: Change $36\frac{1}{5}$ to a decimal.

Step 1. Change $\frac{1}{5}$ to a decimal

$1 \div 5 = .2$

Step 2. The decimal value is now .362

Step 3. Write this as a fraction: $\frac{362}{1000}$

D. Addition and Subtraction of Decimals

When adding decimals the addends and sum must be written such that the decimal points are in one column. Once the decimal points are lined up, proceed as if adding whole numbers. It is frequently helpful to supply zeros after the decimal numbers as place holders. Zeros added to the right of any decimal numbers will not change the value of the number.

Example: Add 2, 7.8, and 3.34

$$\begin{array}{r} 2.00 \\ 7.80 \\ +\ 3.34 \\ \hline 13.14 \end{array}$$

When subtracting decimals, the decimal point must be lined up in the minuend, subtrahend and difference. It is particularly helpful to supply zeros to the right of the decimal numbers in subtraction.

Example: Find the difference when 1.236 is subtracted from 12.88.

$$\begin{array}{r} 12.880 \\ -\ 1.236 \\ \hline 11.644 \end{array}$$

E. Multiplication and Division of Decimals

1. Multiplying and dividing by numbers that are multiples of ten

If a whole or decimal number is multiplied by 10, 100, 1,000, etc., the decimal point in the number is moved the same number of places to the right as there are zeros in the multiplier.

Examples: $13.259 \times 100 = 1325.9$

$6.37 \times 10 = 63.7$

If a whole or decimal number is divided by divisors, such as 10, 100, 1,000, etc., the decimal point in the number is moved the same number of places to the left as there are zeros in the divisor.

Examples: $24.36 \div 10 = 2.436$

$569.2 \div 100 = 5.692$

2. Rules used in multiplying decimals

The decimal points in the multiplicand and the multiplier need not be lined up as they are in addition and subtraction. After the whole-number product is determined, the placement of the decimal point in the product is determined by the combined sum of the number of digits that are to the right of the decimal point in the multiplicand and the multiplier. The number of decimal places in the **product** is equal to the **sum** of the decimal places in the **multiplicand** and the **multiplier**.

Example: Multiply 2.456 by 8.32.

```
    2.456    (3 places to right of decimal point)
  × 8.32     (2 places to right of decimal point)
    4912
    7368
  19648
  2043392    (decimal point is placed 3 + 2 or 5 places
  20.43392   from right)
```

3. Rules used in dividing decimals

In division of decimals, the divisor must be a whole number. If the divisor is not a whole number, **move the decimal point enough places to the right to make it a whole number. To compensate for moving the decimal point in the divisor, the decimal point must be moved the same number of places to the right in the dividend. The** decimal point in the quotient is placed directly above the decimal point of the dividend after the decimal point has been moved.

Example: Divide 5.06688 by 3.248.

```
                1. 56
     3.248 )5.066 88
            3 248
            1 8188
            1 6240
              19488
              19488
```

Exercise 3–1

ADDITION AND SUBTRACTION OF DECIMALS

Write the numerals for the following:

1. Two thousand, forty-five and sixty-seven thousandths.   _____

2. Six and nine hundred thousandths.   _____

3. Ninety-five and one hundred eighty-seven ten-thousandths.   _____

4. One hundred and twenty-one hundredths.   _____

5. Eleven and six thousand, three hundred thirty-four hundred-thousandths.   _____

Express the following numerals in words:

6. 201.206

   _____

7. 16.0215

   _____

8. 20.1252

   _____

9. 11,575.9

   _____

10. 376.02777

    _____

    _____

Express as decimal equivalents, rounding to hundredths place.

11. $\dfrac{21}{50}$ =

12. $\dfrac{19}{45}$ =

13. $\dfrac{7}{85}$ =

14. $\dfrac{38}{125}$ =

15. $\dfrac{25}{9}$ =

Express as common fractions reduced to lowest terms:

16. $0.05\dfrac{1}{4}$ =

17. 0.36 =

18. 32.875 =

19. 1.41 =

20. 0.0025 =

Rewrite the following problems in column form, add, and check by the reverse addend method:

21. $246.2482 + 1.7006 + 0.6825 + 8{,}954.213$ =

22. $39.8 + 25.87 + 70.0002 + 196.8229$ =

Name_____ Date_____

23. 6,432.7 + 836 + 216.003 + 0.206 =

24. 30.026 + 893.21 + 0.835 + 200.992 =

25. 0.0078 + 4.93 + 675.89 + 22.81 =

Rewrite in column form, subtract, and check by addition:

26. 14.682 − 1.008 =

27. 397.13 − 83.7825 =

28. 29.65 − 9.008 =

29. 492.06 − 78.395 =

30. 216.6 − 98.978 =

Add the following quantities:

31. 2.69 kilograms + 0.24 kilogram + 16.234 kilograms.   _____

32. 12.48 square feet + 10.02 square feet + 4.683 square feet.   _____

33. 93.2 hectares + 12.9 hectares + 2.08 hectares.   _____

34. Find the sum of $72.36, $193.97, and $1,006.31.   _____

35. Add 1.34 inches and 2.6009 inches, rounding to thousandths place.   _____

Subtract the following quantities:

36. 11.678 liters − 9.072 liters.   _____

37. 267.03 cubic inches − 200.79 cubic inches.   _____

38. $100.73 − $89.29.   _____

39. Which quantity is greater, 0.340 acre or 0.304 acre?   _____

40. Subtract 2.078 kilograms from 4.390 kilograms, rounding the difference to hundredths place.   _____

Multiply the following:

41. 2.306 × 10.   _____

42. 14.92 × 1000.   _____

43. 23.4 × 100.   _____

44. 47.235 × 10.   _____

Divide the following:

45. 56.3 ÷ 10.   _____

46. 258.49 ÷ 1000.   _____

47. 375.342 ÷ 100.   _____

48. 256.35 ÷ 10.   _____

Solve the following problems:

49. On three consecutive days a weather station recorded the following amounts of rainfall: .59 inch, 1.01 inch, and .2 inch. What was the total rainfall in the three-day period?

_____

50. A bushel of grain harvested from a rice crop removes 2.82 kilograms of nitrogen from the soil. If the straw that is a by-product of a bushel of grain removes an additional 1.73 kilograms of nitrogen, what total amount of nitrogen is removed by the rice plant in producing a bushel of rice?

_____

51. Gasohol for non-highway use costs $1.24 per gallon. How much cheaper is a gallon of diesel fuel at $1.15 per gallon?

_____

52. A cattle farmer purchased a liter of pour-on parasiticide at $128.90, two fly blocks at $7.50 each and one bag of Hi-mag salt at $9.95. What was the total amount of the purchase?

_____

53. A custom operator recorded the tractor's hour meter reading as 1,645.6 at the beginning of a plowing job. When the job was completed, the reading was 1,738.4. How many hours of engine operating time had been spent plowing?

_____

54. The section of sheep fence shown in Figure 3–1 has line wires spaced as shown. What is the height of the fence?

_____

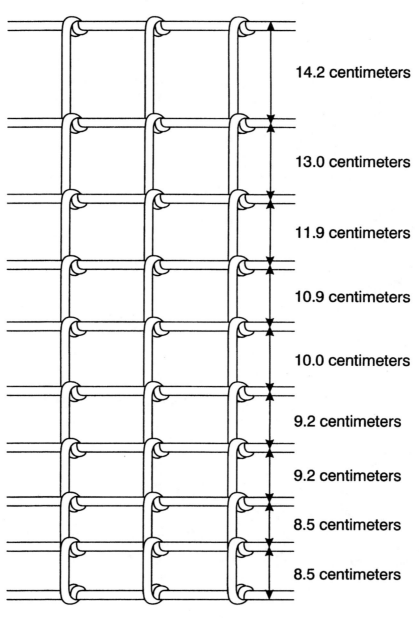

14.2 centimeters

13.0 centimeters

11.9 centimeters

10.9 centimeters

10.0 centimeters

9.2 centimeters

9.2 centimeters

8.5 centimeters

8.5 centimeters

Figure 3–1

55. Cropping records show that a corn producer harvested 89,049.6 bushels of shelled corn from 576 acres of cropland. Last year 75,801.9 bushels of equal quality corn were harvested from the same acreage. How much larger is this year's crop?

_____

56. Analysis of a farmer's records shows the following costs related to producing an acre of sun-flowers: seed, $13.25; fertilizer, $16.50; chemicals, $11.10; fuel, $9.86; repairs and maintenance of equipment, $11.20; and land taxes $9.19. What is the total cost?

_____

57. Figure 3–2 shows a cross-sectional view of a tower silo and the footing that supports it. From the information shown, find the outside diameter of the footing.

_____

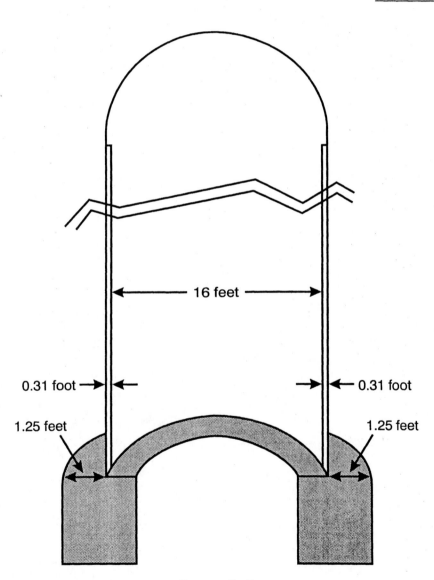

Figure 3–2

58. Sales of fresh farm-raised catfish totaled 9.51 million pounds in April. Frozen catfish sales totaled 14.0 million pounds for the same month. How many million pounds more of frozen fish were sold than of fresh fish?

_____

59. Standard weight of a bushel of 15.5 percent moisture shelled corn is 56 pounds. If it takes 3.64 additional pounds of 20.5 percent shelled corn to be equivalent to a bushel of 15.5 percent corn, what weight of 20.5 percent corn is equivalent to the 56 pounds per bushel of 15.5 percent corn?

_____

60. A 450-kilogram steer gaining 1.4 kilograms per day is fed corn silage, corn and cob meal, and soybean meal. If the silage furnishes 0.22 kilogram of protein, corn and cob meal 0.42 kilogram, and soybean meal 0.41 kilogram, what is the total amount of protein fed the steer?

_____

61. The May all wheat price, at $2.58 per bushel, was down 48 cents below May of the previous year. What was the price the previous year?

_____

62. The following advertisement appeared in an agricultural supply company's sale catalog:

> Treated fence posts. 4 × 4 × 8. Reduced $0.47 per post when purchased in quantities over 100 posts.

A rancher purchases 230 posts at $6.25 each. How much did the rancher pay for the posts?

_____

How much more would the rancher have paid if the posts had not been on sale?

_____

63. Figure 3–3 shows field notes of a survey crew determining the difference in elevation from point A to point B. An engineer's level is set up at point 0. The backsight and foresight readings on the surveying rods indicate the height of the instrument above points A and B, respectively. Find the difference in elevation.

_____

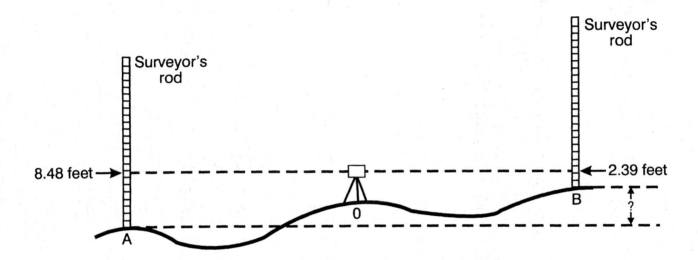

Figure 3–3

64. Average normal body temperature of swine is 102.6 degrees Fahrenheit. During an outbreak of erysipelas, a pig had a temperature of 105.3 degrees Fahrenheit. How much above normal was the pig's temperature?

_____

65. The check register of a farmer's checkbook indicates a total of $11,678.95 in the account. After paying a bill of $8,341.68 for herbicide and fertilizer, what was the amount in the account?

_____

66. Butter prices received for 68-pound boxes meeting USDA Grade AA standards averaged 98.7 cents per pound for a recent week. The price per pound decreased 5.5 cents per pound from the previous week. What was the price the previous week?

_____

67. A 400-kilogram, 18-month-old colt is fed a ration of timothy hay and oats. The recommended daily feed intake (dry matter basis—abbreviated DM) is 6.24 kilograms. If 4.46 kilograms (DM) of timothy hay are fed, what amount of oats (DM) are needed?

_____

Exercise 3–2

MULTIPLICATION AND DIVISION OF DECIMALS

Perform the indicated operations:

1. 765 ÷ 10 = _____

2. 2.936 × 100 = __293.6__

3. 489.6 × 1,000 = _____

4. 92.3 ÷ 100 = _____

5. 670.3 ÷ 1,000 = _____

6. 20.045 × 10 = _____

In the numeral 5,783.39954, round off to:

7. (hundredths)_____

8. (tenths) _____

9. (hundreds) _____

10. (thousandths)_____

11. (tens)_____

12. (ten-thousandths) _____

Multiply the following decimals and check by the reverse factor method:

13. 572.9   Check
    4.006

14. 954.3   Check
    2.248

15. 6.835   Check
    0.345

16. 299.7   Check
    67.32

Divide and express the quotients to the nearest hundredth:

17. 2.7835 ÷ 4.63 =   18. 0.9862 ÷ 0.629 =

19. 673.4 ÷ 52.1 =   20. 10.02658 ÷ 49.61 =

21. 1.48 × 6.783 pounds = _____ pounds. (Round to tenths.)

22. 2.683 meters × 4.82 meters = _____ square meters. (Round to thousandths.)

23. 7.52 feet × 3.91 feet × 5.06 feet = _____ cubic feet. (Round to hundredths.)

24. 46.9 liters per hectare × 16.2 hectares = _____ liters. (Round to tenths.)

25. 14.67 inches × 1.06 inches = _____ square inches. (Round to tenths.)

26. 76.54 acres ÷ 4.06 acres per gallon = _____ gallons. (Round to tenths.)

27. 272.4 square meters ÷ 100 square meters per hectare = _____ hectares. (Round to tenths.)

28. 247.8 square yards ÷ 3.2 yards = _____ yards. (Round to tenths.)

If the products or quotients in problems 29 through 50 have more than two decimal places, round to two decimal places.

29. The custom rate for sugarbeet thinning is $23.67 per acre. What would it cost to have 245.5 acres of beets thinned?

_____

30. A semi's odometer indicates the truck has traveled 62,440 miles during the past year transporting grain. If depreciation on the truck is charged at the rate of $0.13 per mile traveled, what amount of depreciation should be charged for the year?

_____

_____

72

31. A desirable beef animal should yield 0.64 kilogram of carcass for each day of age. How old should a desirable beef animal be at time of slaughter if its carcass weighs 297.6 kilograms?

_____

32. Production records for an apiary show that each colony of bees produces an average of 102 pounds of honey. If honey sells for 49.5 cents per pound, what is the value of the honey from a bee colony?

_____

33. A grade of seed corn has 1,465 seeds per pound. If a "unit" of seed corn contains 80,000 seeds, approximately how many pounds does a unit of this corn weigh? (Round to one-tenth of a pound.)

_____

34. In addition to the nutrients needed for body maintenance, a dairy cow requires 0.048 kilogram of digestible protein for each kilogram of 3.5 percent milk produced per day. What are the daily additional nutrient requirements for a cow producing 32.4 kilograms of 3.5 percent milk per day?

_____

35. A vineyard produced 614.6 metric tons of raisin variety grapes. If it takes 4.05 metric tons of grapes to produce 1 ton of raisins, what amount of raisins could be processed from the vineyard's grapes?

_____

36. Field efficiency of a tractor is maximized when the horsepower of the tractor is matched with the power requirements of the operation to be done. What width of field cultivator should be matched with the tractor having the horsepower given in Figure 3–4 and working under the indicated soil conditions?

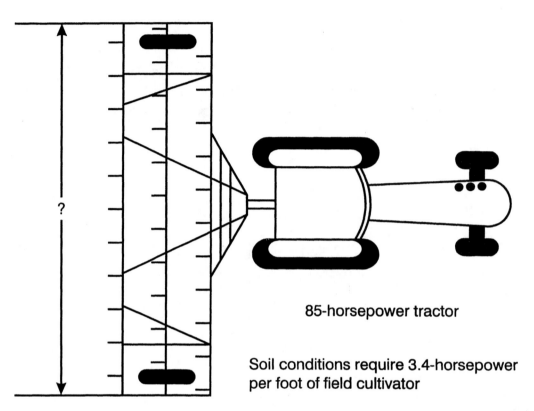

85-horsepower tractor

Soil conditions require 3.4-horsepower per foot of field cultivator

Figure 3–4

In many practical applications, it is necessary to compute the relationship between two units. To compute relationships, such as cost per ounce or miles per gallon, read the "per" as a division sign.

Examples: A truck uses 11.4 gallons of gasoline to travel 206 miles. How many miles per gallon does the truck average?

206 ÷ 11.4 = 18.07 miles per gallon

A 25-pound bag of feed costs $10.97. What is the cost per pound?

$10.97 ÷ 25 = $0.44 per pound

37. If 8.5 fluid ounces of a pour-on parasiticide costs $49.20. What is the cost per ounce?

_____

38. If a farmer can purchase 33.8 fluid ounces of the same pour-on parasiticide at $129.00, how much is saved per ounce by buying the larger bottle?

_____

39. A truck hauling lumber travels 162 miles and uses 12.2 gallons of diesel fuel. Compute the average miles per gallon.

_____

If diesel costs $1.15 per gallon, what is the cost for the trip?

_____

40. A 50-pound block of protein supplement for cattle costs $16.20, a 200-pound block costs $31.00, and a 250-pound block costs $35.76. If a rancher needs 600 pounds of protein supplement, what will be the most cost-efficient way to purchase the supplements? Explain your answer.

_____

_____

41. A horse is exercised by the motorized "hot walker" shown in Figure 3–5. Use the formula for the circumference of a circle, C = πD, and the data shown to find the distance walked by a horse making 80 rounds in the walker. (Use 3.14 for π.)

_____

4.6 meters

Figure 3–5

42. What amount of fuel is needed to harvest 95 acres of wheat if a grain combine uses 1.4 gallons of diesel fuel per acre?

_____

43. A carton of 500 basswood section boxes for producing comb honey sells for $72.50. What is the cost per section box?

_____

Name_____ Date_____

44. Maple syrup is 1.33 times as heavy as water. If a gallon of water weighs 8.34 pounds, what is the weight of a gallon of maple syrup?

_____

45. If producing 60 bushels of barley per acre costs a farmer $153.10, what price must be received per bushel to break even?

_____

46. A large round bale of hay is wrapped with twine as shown in Figure 3–6.

a. If a ball of twine contains 16,000 feet of twine, approximately how many bales can be made with one ball of twine?

_____

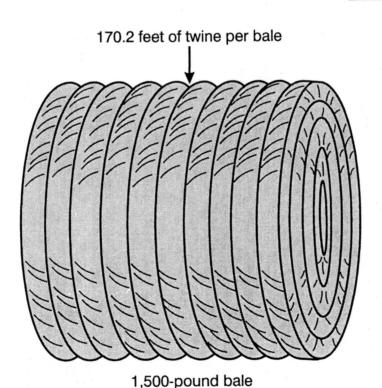

170.2 feet of twine per bale

1,500-pound bale

Figure 3–6

b. If a ball of twine used for large round bales costs $21.95, what is the cost of twine per bale?

_____

47. A farmer hired a custom operator to plant corn with an eight-row, 30-inch-row corn planter. What is the cost per acre if the hourly rate is $89.28 and 7.64 acres can be planted per hour?

_____

48. A soil conservation technician is assisting a farmer in making plans for a contour strip cropping system on rolling land, as shown in Figure 3–7. The first year, only two strips will be used for row crops, but crop rotation on the strips requires that all strips be of appropriate width to accommodate row-crop equipment. The farmer wants to make four rounds, as indicated in the diagram, with six-row equipment having 2.5-foot row spacing. What strip width is best for the farmer's machinery?

_____

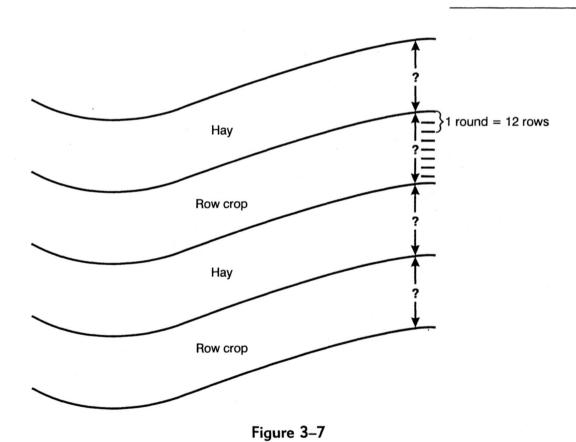

Figure 3–7

Name_____ Date_____

Determine Plant Spacing:

Example: Corn population is 20,000 plants per acre. If rows are 36 inches apart, how far apart must the seed be spaced?

Step 1. Determine the linear feet of row per acre.

Square feet of one acre ÷ distance between rows (in feet)

43,560 ÷ 3 = 14,520 linear feet

Step 2. Determine linear inches of row per acre.

Linear feet × 12 inches per foot

14,520 × 12 = 174,240 linear inches

Step 3. Determine distance of seeds in inches.

Linear inches ÷ desired plant population per acre

174,240 ÷ 20,000 = 8.7 inches

Seeds should be spaced 8.7 inches apart.

49. If soybean population is 22,000 seeds per acre and rows are 36 inches apart, how far apart do seeds need to be spaced?

_____

50. If plant population is 18,000 seeds per acre and rows are 24 inches apart, how far apart do seeds need to be spaced?

Name_____ Date_____

Exercise 3–3

COMBINED OPERATIONS WITH DECIMALS

Solve problems 1 through 10, rounding as indicated.

1. $16.304 + 2.58 - 11.003$ (Round to tenths.)   _____

2. $2.63 \times 4.8 \div 14.1$ (Round to thousandths.)   _____

3. $2.48 + 5.91 \times 3.2 - 16.02$ (Round to hundredths.)   _____

4. $2.25 \div [(2.83 - 1.42) \times 6.3]$ (Round to tenths.)   _____

5. $0.60[0.002 + (0.96 - 0.32) \times 1.05] \div 0.16$ (Round to hundredths.)   _____

6. 4.9 centimeters (14.2 centimeters + 8.05 centimeters)
   = _____ square centimeters. (Round to hundredths.)   _____

7. 4.3(7.35 feet) + 16.83 feet = _____ feet.
   (Round to tenths.)   _____

8. (1,029.63 square rods + 206.78 square rods) ÷ 160 square
   rods per acre = _____ acres. (Round to tenths.)   _____

9. 16.3 liters + 29.3 liters × 12.2 = _____ liters.
   (Round to tenths.)   _____

10. 25.4 bushels + 3.2(64.3 bushels − 15.9 bushels)
    = _____ bushels. (Round to tenths.)   _____

81

If computations for problems 11 through 25 result in numbers having more than two decimal places, round to two decimal places.

11. A gasoline-powered corn combine uses 2.25 gallons of gasoline per acre. A comparable diesel-powered combine consumes 1.6 gallons of fuel. With gasoline costing $1.24 per gallon and diesel fuel $1.15 per gallon, what is the difference in fuel costs in harvesting 240 acres of corn?

_____

12. Figure 3–8 shows a classified ad that appeared in a local newspaper. A farmer responds to the ad and agrees to the advertised price and the seller's estimate of 3.9 tons per foot of settled silage. What is the total cost of the silage?

_____

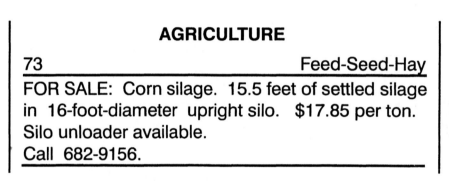

AGRICULTURE	
73	Feed-Seed-Hay

FOR SALE: Corn silage. 15.5 feet of settled silage in 16-foot-diameter upright silo. $17.85 per ton. Silo unloader available.
Call 682-9156.

Figure 3–8

13. A hay baler for large round bales can be purchased for a cash price of $15,995. An installment plan is also available with terms of $1,000 down and $362.80 per month for 48 months. What is the additional cost for the installment?

_____

14. A flock of 145 ewes was shorn with an average fleece weight of 8.7 pounds of wool. If the farmer received $782.13 for the wool, what was the market price of a pound of wool?

_____

Name_____ Date _____

15. A 300-foot roll of galvanized wire fence, 3 feet high, weighs 162 pounds. Find the total number of feet of fence contained in two partially used rolls weighing 54.1 pounds and 33.4 pounds.

_____

16. A farmer purchased 4 tons of baled hay for $366.40. If the bales average 800 pounds, what is the price per bale of hay?

_____

17. Three peanut research plots are located on rental property. If the plots contain 1.25 acres, 0.73 acres, and 2.4 acres, and rent is based on $253 per acre, what is the rental charge?

_____

18. The table below shows a portion of a hog grower's record of feeder pig purchase. Find (a) the total amount spent for pigs in January and (b) the average price paid per pig.

a._____

b._____

Date of purchase	Number of head	Cost per pig	Total cost
January 2	156	$32.50	$_____
January 9	168	35.60	_____
January 17	147	42.80	_____
January 25	154	38.25	_____
		Total cost (January)	$_____

19. If the grain storage rate is 6.9 cents per bushel per month in a certain area, what will it cost a corn producer to store 4,450 bushels of corn from November 1 through July 31?

_____

20. A fence is to be built for a sheep enclosure. To find the approximate length of fence required, the rancher paces off the lengths of the three sides to be fenced. Notes on the procedure are shown in Figure 3–9. If the rancher's paces average 0.86 meter, what length of fence is needed?

_____

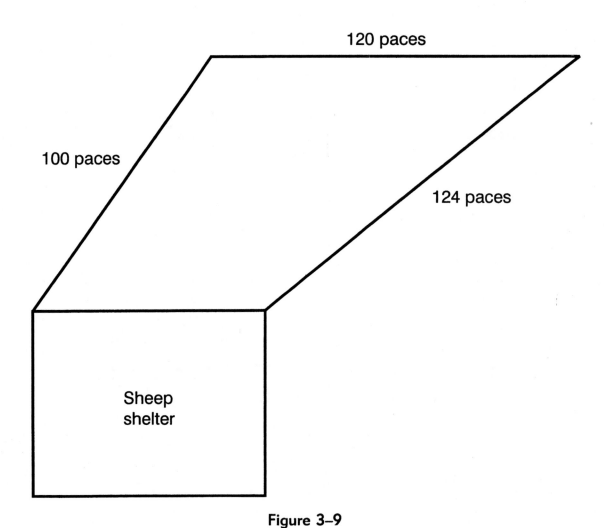

120 paces

100 paces

124 paces

Sheep
shelter

Figure 3–9

21. A commercial apiary sells its entire honey crop in 1-liter glass jars. If their 362 colonies produced 54.4 kilograms of honey per colony, how many completely filled jars can be sold? (Honey weighs 1.41 kilograms per liter.)

_____

22. During the gestation period, a flock of 168 ewes is fed an average of 9 ounces of shelled corn per head per day.

   a. How many pounds of corn are fed to the flock per day?   _____

   b. If shelled corn weighs 56 pounds per bushel, how many bushels of corn are fed during the month of November?   _____

   c. If the price of corn averaged $2.34 per bushel during November, what was the cost of the corn fed to the flock that month?   _____

23. A 2¢-per-gallon discount is received if payment is made for bulk fuel within 10 days of delivery. If 475 gallons of $1.16 diesel fuel are delivered to a farmer's storage tank, what amount needs to be paid for the fuel if the discount is received?

_____

24. The table below shows the average monthly prices received by farmers for soybeans in three consecutive years.

   a. The calendar year having the greatest range in price was year _____.

   b. Comparing the prices of year B with year C, during which month was the difference in price the greatest? _____

   c. The greatest month-to-month change in soybean prices occurred between _____ (month) and _____ (month) of year _____.

d. The average price received for soybeans during the first half of year B was

_____ .

e. The average price received during April of the three years was _____ .

Year	Jan	Feb	Mar	Apr	May	Jun	Jul	Aug	Sep	Oct	Nov	Dec
A	4.46	4.50	4.46	4.52	4.87	6.16	6.73	6.07	6.65	5.90	6.11	6.56
B	6.81	7.06	7.83	9.05	9.24	8.13	6.52	5.48	5.17	5.28	5.61	5.69
C	5.75	5.53	6.20	6.49	6.77	6.69	6.39	6.21	6.19	6.26	6.39	6.49

(Data given in dollars per bushel.)

25. The fuel storage tank on a large diesel tractor has a capacity of 95 gallons.

a. What is the value of the tankful of fuel if diesel fuel costs
$1.15 per gallon?   _____

b. Diesel fuel weighs 6.9 pounds per gallon. What additional
weight does a full tank of fuel give the tractor?   _____

c. If a plowing operation requires 6.7 gallons of fuel per hour,
how long can the tractor be operated without refueling?   _____

d. If tillage equipment can cover 8.26 acres per hour, using
3.7 gallons of fuel per hour, how many acres can be tilled
on a tankful of fuel?   _____

Exercise 3–4

APPLICATION EXTENSION

Calculating a Landscape Cost Estimate

A **cost estimate** is an itemized breakdown of the expenses involved in developing a landscape. It is important and useful for the landscaper as well as the client. The client should know the total cost of the design before the landscape work begins. The landscaper must know what he or she must charge in order to make a profit and remain in business.

Pricing a Plan

Solve the following:

1. Plant Materials

   3 silver maples, 5' B&B at $7.00 each.   _____

   2 redbuds, 5' B&B at $9.00 each.   _____

   20 deciduous goldenbell forsythia, 2' BR at $4.00 each.   _____

   20 hybrid azaleas, 1-gallon containers at $4.50 each.   _____

   3–4' pyracanthas, 1-gallon containers at $5.50 each.   _____

   250 English ivy cuttings at $.31 each.   _____

   2 sycamores, 10–12' B&B at $25.00 each.   _____

2. Construction Materials

   100 feet of redwood rail fence at $9.50 / 10 feet.   _____

   300 blocks of concrete pads for patio at $.75 each.   _____

   12 tons pine bark at $20.00 per ton.   _____

   6 rolls black plastic at $22.00 per roll.   _____

   3 outdoor lights at $35.00.   _____

   Figure the installation cost by taking ½ the value of construction materials.   _____

3. Turf Grass

   10,000 square feet sodded with zoysia at $.12 per square foot. _____

   12 square feet seeded with Bermuda at $.08 per square foot. _____

   Total _____

   Note: These are wholesale prices. In this method of determining cost, the exact cost is determined first. Then to cover overhead, management, and profit, 30 percent of the total cost is added.

2. Given the following cost and information, what would be the charge for this landscape job? _____

   #1 = 1 gallon = $2.00
   #3 = 3 gallon = $5.00
   #5 = 5 gallon = $8.00

Plant Material

   10 dwarf yaupon holly, #1 _____

   5 Japanese cleyero, #3 _____

   20 blue rug juniper, #1 _____

   10 dwarf Buford holly, #3 _____

   3 fosteri holly, #5 _____

   2 sycamore trees, 10–12' @ $25.00 each. _____

Construction Materials

   3 rolls black plastic @ $15.00/roll. _____

   3 tons pine bark @ $15.00/ton. _____

   100 hours labor @ $8.00/hour. _____

   Design cost $80.00 _____

   Transportation and equipment costs $400.00 _____

   Total Charged to Customer _____

4

PERCENT AND PERCENTAGES

Key Terms

assessed valuation
banker's interest
base
cash discount
commission
compound interest
converting markups
cost value
date of note
E.O.M. dating
exact interest
face of note

find cost
find selling price
germination percentage
graduated commission
interest
markup as a percent
maturity date
maturity value
ordinary dating
ordinary interest
percent
principal of note

promissory note
property tax
rate of interest
retail value
R.O.G. dating
salary and commission
sales tax
simple interest
straight commission
tax assessor
tax rate
time of note

Career Connections

AGRICULTURAL ECONOMIST

Agricultural economists study how agriculture operates in the economic system. They collect information on trends that relate to agriculture, analyze current situations, and predict future supply and demand to help producers make decisions.

Agricultural economists need college degrees in agricultural economics or a related area. Practical experience in agriculture is important. Most agricultural economists work for government agencies, agribusinesses, and universities.

A. What Is Percent?

Percent is a fraction with 100 as the denominator. The percent symbol (%) is placed to the right of the numerator and replaces the denominator of 100.

Example: $\frac{24}{100} = 24\%$.

B. Basic Rules for Percent

1. **Changing a decimal to a percent.** Move the decimal point two places to the **right** (equivalent to multiplying by 100) and attach the percent symbol to the **right** side of the number.

   Example: Change 0.35 to a percent.

   $$0.35 = 35\%$$

   $$.915 = 91.5\%$$

2. **Changing a percent to a decimal.** Remove the percent sign and move the decimal point two places to the **left**.

   Examples: Change 28% to a decimal.

   $$28\% = 0.28.$$

   $$352\% = 3.52$$

   Note: Mathematical operations such as multiplication and division cannot be performed with the percent sign attached to a number. The percent sign must be removed and the number changed into a decimal or a fraction.

3. **Changing a fraction to a percent.** The fraction is first changed to a decimal by dividing the numerator by the denominator. The decimal is then changed to a percent.

   Examples: Change $\frac{4}{5}$ to a percent.

   $$\frac{4}{5} = 0.80 \times 100 = 80\%$$

   $$\frac{3}{8} = .375 \times 100 = 37.5\%$$

4. **Changing percents to common fractions.** Remove the percent sign and move the decimal two places to the left. Change the decimal to a common fraction having a denominator of 100. Reduce the fraction to lowest terms.

Example: Change 24% to a fraction.

$$24\% = 0.24 = \frac{24}{100} = \frac{6}{25}.$$

Example: Change 102% to a fraction.

$$102\% = 1.02 = \frac{102}{100} = \frac{51}{50}.$$

Example: Change $16\frac{2}{3}\%$ to a common fraction.

$$16\frac{2}{3}\% = 0.16\frac{2}{3} = \frac{\frac{50}{3}}{100} = \frac{50}{3} \times \frac{1}{100} = \frac{1}{6}.$$

C. Solving Problems Using Percent

There are two equivalent methods for solving problems using percent.

1. Using number sentences with percents

Many percent problems can be stated in an expression like "What percent of 96 is 24?". Making a number sentence out of the expression and inserting the proper arithmetic symbols the expression becomes N% $\times$ 96 = 24. The problem now is an operation involving multiplication; that is, a factor times a factor yields a product. The mathematical operation of division is the reverse of multiplication, so the product divided by the known factor yields the other factor. Solving the problem, 24 $\div$ 96 = 0.25 $\times$ 100 = 25%.

Rules for number sentences to solve percent problems

1. If the product and one factor are known, divide the product by the known factor to find the other factor.

2. If both factors are known, simply multiply the factors together to find the product.

2. Using the percent formula to solve percent problems

Percent is related to two other quantities, **base** and **percentage**. When we find percent of a number, the answer is called a **base**. The part of the base is the **percentage**. The percent **(rate)** represents the relationship between the part (percentage) and the whole (base). From this relationship we get the formula:

$$R = \frac{P}{B},$$

which can be used to find the percent when the percentage and the base are known. When the base and the percent are known, the formula is revised to give the percentage by the formula: $P = B \times R$.

When the percentage and the rate are given, the base can be found by using the formula in the form of:

$$B = \frac{P}{R}.$$

Guidelines for using the formula to solve percent problems

1. Identify the known and unknown quantities as the base, percentage, and percent.

2. Choose the formula that presents the correct relationship.

3. Solve.

3. Examples showing solution of problems by both methods

Problem: The market price of soybeans was quoted at $7.20. One week later the price was $8.28. What was the percent increase in price?

Number Sentence:

$1.08 (increase) is what % of $7.20?
$1.08 = N% × $7.20
$1.08 ÷ $7.20 = 0.15 × 100 = 15%

Formula:

Rate = ?, Base = $7.20, Percentage = $1.08 (increase)

$$R = \frac{P}{B}, \ R = \frac{\$1.08}{\$7.20} = 0.15 \times 100 = 15\%.$$

Problem: A $3,245.00 purchase was made in a state where the sales tax was 5 percent. What amount of sales tax should be paid?

Number Sentence:

5% of $3,245.00 is N.
0.05 × $3,245 = N.
0.05 × $3,245 = $162.25.

Formula:

B = $3,245.00, P = ?, R = 5%
P = B × R
P = $3,245.00 × 0.05 = $162.25.

Problem: A salesperson makes 16 percent commission on all sales. What was the total amount of sales if the commission earned was $270.00?

Number Sentence:

16% of what amount is $270.00?
0.16 × N = $270.00
$270.00 ÷ 0.16 = $1,687.50.

Formula:

B = ? P = $270.00, R = 16%
$B = \dfrac{P}{R}$, $B = \dfrac{\$270.00}{0.16} = \$1,687.50.$

Problem: A horse feed consists of cracked corn and crimped oats. If the mixture is 80 percent oats and 8 kilograms of corn are in each bag, what is the weight of a bag of feed?

Note: In this problem we must first examine the given information to see if the percent and percentage relate correctly to the base. The percent of the corn is 100% — 80% or 20%.

Number Sentence:

8 kg is 20% of N.
8 kg = 0.20 × N
8 kg ÷ 0.20 = 40 kg.

Formula:

B = ?, P = 8 kg, R = 20%
$B = \dfrac{P}{R}$, $B = \dfrac{8 \text{ kg}}{0.20} = 40 \text{ kg.}$

D. Percents Greater Than 100 Percent

All percentage relationships present a comparison of one quantity to another. When the quantity "to be compared" is larger than the quantity "compared to," the result is a percent greater than 100 percent. This type of problem is difficult to set up for solution by formula but is easily solved using the number sentence.

Example: Forty pounds is what percent of 20 pounds?

$$40\# = N\% \times 20\#$$
$$40 \div 20 = 2.00 \times 100 = 200\%.$$

Example: What number is 160% of 24?

$$N = 1.60 \times 24$$
$$N = 38.4.$$

Example: What number is 150 percent greater than 15?

150% greater than 15 is 250% of 15.

$$N = 2.50 \times 15$$
$$N = 37.5.$$

Example: The cost of an article (including sales tax of 4 percent) is $15.60. What is the marked price of the article?

$15.60 is 104% of N (marked price).

$$\$15.60 = 1.04 \times N$$
$$\$15.60 \div 1.04 = \$15.00.$$

Exercise 4–1

BASIC PERCENTAGE PROBLEMS: RELATIONSHIP OF COMMON FRACTIONS, DECIMALS, AND PERCENTS

Solve the following problems. If results of calculations have more than two decimal places, round to two decimal places. Round percents to the nearest tenth of a percent.

Supply the missing equivalents in the following table:

	Fraction	Decimal	Percent
1.	2/5	0.4	40 %
2.	1/8	0.125	12.5%
3.		0.025	
4.	1/2,000		
5.			0.006%
6.			16⅔%
7.	21/400	0.525	5.25 %
8.		0.3125	
9.			⅕%
10.	3/50		
11.		1.25	
12.			137.5%
13.	1⅘		
14.			800%
15.		1.83⅓	
16.			666⅔%
17.	1¹³⁄₂₅		
18.		1.44	
19.	2⁵⁄₁₆		
20.			525%

Complete the following mathematical sentences.

21. 15 is _____% of 25.

22. 112 is 14% of _____.

23. ½ is 80% of _____.

24. 90% of 20.4 is _____.

25. 9 is _____% of 72.

26. 300% of _____ is 18.

27. 102% of 16 is _____.

28. 25% more than 14 is _____.

29. 42 is _____% of 35.

30. 224% of 60 is _____.

Solve the following problems.

31. Horse feed sales at a feed supply store amounted to $1,085.14 this week. The manager says this amount is 15 percent greater than the last week's sales. What was the amount of last week's horse feed sales?

_____

32. Many field operations become increasingly efficient as the length of the field increases. If a tillage operation is 130 percent more efficient on field A than on field B, how many acres can be covered per hour on field A if 3.2 acres can be tilled per hour on field B?

_____

33. Cottonseed meal having a 33.5 percent digestible protein content is to be fed as a supplement for beef steers on a ration of grass silage and ground ear corn. If the supplement needs to furnish each steer with 0.30 kilogram of digestible protein, what amount of cottonseed meal should be fed?

_____

34. A yield check on an alfalfa-brome hay crop indicates a potential yield of 1,887 kilograms of dry matter per hectare. Field operations are responsible for the following losses: cutting and conditioning, 3.5 percent; raking, 4.6 percent; and baling, 6.3 percent. If an inch of rain caused an additional 3.4 percent loss of dry matter, what amount of hay (dry matter) was harvested?

_____

35. The use of irrigation on droughty soil increased the average yield of sunflowers from 1,432 kilograms per hectare to 2,934 kilograms per hectare. What percent yield increase can be attributed to the irrigation system, assuming all other growing and weather conditions were the same?

_____

36. The pod-filled soybean plants shown in Figure 4–1 are harvested, leaving the stubble length shown. If the potential yield of soybeans is 49 bushels per acre, what is the amount of loss per acre due to stubble?

_____

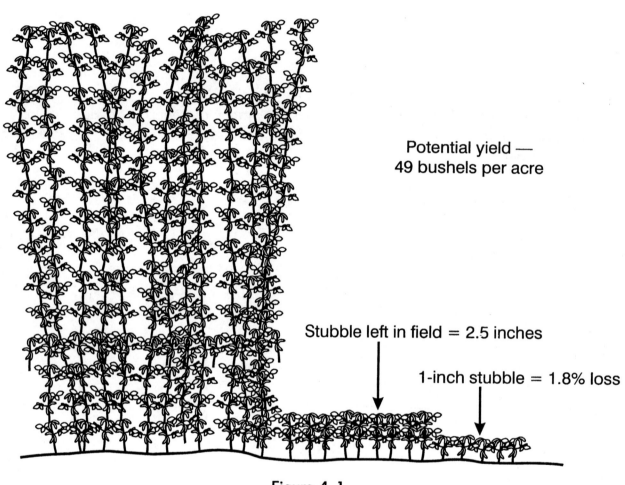

Potential yield —
49 bushels per acre

Stubble left in field = 2.5 inches

1-inch stubble = 1.8% loss

Figure 4–1

37. In making an analysis of alfalfa seed, a technician found 6.942 grams of pure seed and 0.178 gram of weed seed. What is the percent of pure seed in the sample?

_____

38. An acreage of soybeans is planted and harvested by three young farmers wanting to earn money for college. Because of different amounts of time and equipment furnished, the first farmer is to get 35 percent of the profit, the second 37.5 percent, and the third the remainder of the profit. If the net income is $15,476, what amount does the third farmer get?

_____

39. During last year's lactation, a dairy cow produced 10,043 kilograms of milk testing 4.2 percent butterfat. What was her butterfat record for the year?

_____

40. A corn planter is set to plant 60,290 corn kernels per hectare. This planter setting represents a seeding rate 10 percent higher than the desired growing plant population.

a. What is the desired growing plant population?   _____

b. If an 80,000-seed bag of seed corn costs $58.60, what is
the seed corn cost per hectare?   _____

41. A florist has orders for flower arrangements requiring 9 dozen carnations. If the florist has 50 carnations on hand, what percent of the required flowers will need to be ordered from a wholesaler?

_____

42. A self-propelled swather is designed to have a field capacity of 2.38 hectares per hour. If the average cut is shown in Figure 4–2, (a) what percent of a full cut is made and (b) what is the actual field capacity?

a._____

b._____

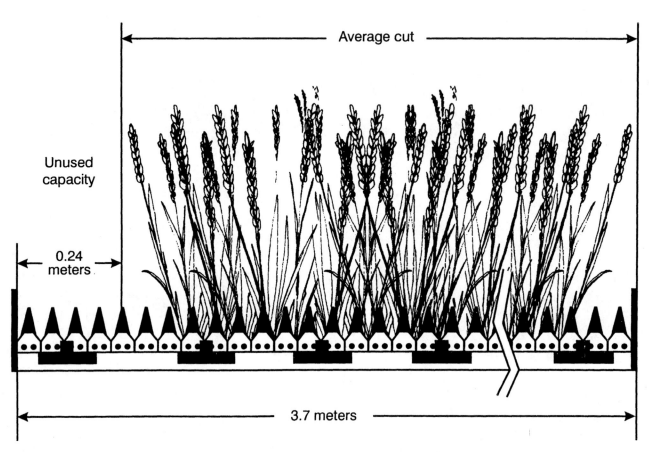

Average cut

Unused capacity

0.24 meters

3.7 meters

Figure 4–2

43. A grower receives 13.6 cents per pound for new potatoes. By the time the potatoes reach the consumer market, the price has increased to 59 cents per pound. Find the percent increase in price from grower to consumer.

_____

44. A 6.5-pound roasting chicken yields a 4.6-pound ready-to-cook carcass. Find the percent loss from dressing and drawing.

_____

PART II. MERCHANDISING AND RETAILING, CASH DISCOUNTS

A. Merchandising and Retailing

Operation of a retail store is one of the most challenging jobs in the business world. It takes real managerial talent to perform all of the functions of merchandising. The successful retailer must know what, when, and where to buy; how to price merchandise for profit; and how to analyze sales trends. Merchandise has a **cost value** and a **retail value**. The difference between these values is called by different names under various circumstances. It is sometimes called margin, gross profit, markon, and markup. All these terms have essentially the same meaning. The term markup will generally be used in this section. Expressed as equations, the relationships between the cost, markup, and retail price are:

$$\text{Cost} + \text{Markup} = \text{Retail} \quad \text{or} \quad C + M = R,$$

$$\text{Retail} - \text{Cost} = \text{Markup} \quad \text{or} \quad R - C = M,$$

$$\text{Retail} - \text{Markup} = \text{Cost} \quad \text{or} \quad R - M = C.$$

1. **Markup as a percent.** The amount of the markup is usually expressed in terms of dollars and cents but is often referred to as either a percent of the cost or a percent of the retail price of the product.

$$\text{Markup as a percent of cost} = \frac{\text{Markup}}{\text{Cost}}.$$

$$\text{Markup as a percent of retail} = \frac{\text{Markup}}{\text{Retail}}.$$

Example: What is the percent of markup on an item that costs 72 cents if the item sells for $1.00?

$$\text{Markup based on cost} = \frac{\text{Markup}}{\text{Cost}} = \frac{28}{72} = 0.389 = 38.9\%.$$

$$\text{Markup based on retail} = \frac{\text{Markup}}{\text{Retail}} = \frac{28}{100} = 0.28 = 28\%.$$

2. **Converting markups.** It is sometimes necessary to convert the markup percent of cost to a markup percent of retail or vice versa. To find the markup percent of cost when given the markup percent of retail, use the following relationship:

$$\text{Markup percent of cost} = \frac{\text{Markup percent of retail}}{\text{Cost percent of retail}}.$$

Example: What is the percent markup based on cost when the markup is 20 percent of retail?

$$\text{Markup based on cost} = \frac{20\%}{80\%} = 0.25 = 25\%.$$

To find the markup percent of retail when given the markup percent of cost, use the following relationship:

$$\text{Markup percent of retail} = \frac{\text{Markup percent of cost}}{\text{Retail percent of cost}}.$$

Example: What is the percent markup based on retail when the markup is 20 percent of cost?

$$\text{Markup based on retail} = \frac{20\%}{100\% + 20\%} = \frac{0.20}{1.20} = 0.166\ldots \text{ or } 16\tfrac{2}{3}\%.$$

3. **Given the cost and the markup percent of cost, find the retail or selling price.**

   a. Multiply the cost by the percent markup to find markup.

   b. Add the markup to the cost.

   Example: African violets cost $2.45 each and are to have a markup of 60 percent of cost. Find the retail price.

$$\text{Cost} + \text{Markup} = \text{Retail}$$
$$\$2.45 + (0.60 \times \$2.45) = \text{Retail}$$
$$\$2.45 + \$1.47 = \text{Retail}$$
$$\$3.92 = \text{Retail}.$$

4. **Given the retail price and the markup percent of cost, find the cost.**

   a. Let the cost equal 100 percent.

   b. Add the markup percent to 100 percent.

   c. Divide the retail price by the total percent.

   Example: A store wants to sell containers of garden pesticide at retail for $1.77 each. The store needs a markup of 65 percent of cost. What can the buyer for the store afford to pay the supplier for the pesticide?

$$\text{Cost} + \text{Markup} = \text{Retail}$$
$$100\%(\text{Cost}) + 65\%(\text{Cost}) = \$1.77$$
$$165\%(\text{Cost}) = \$1.77$$
$$\text{Cost} = \$1.77/1.65$$
$$\text{Cost} = \$1.07.$$

5. **Given the markup and the markup percent of cost, find the cost and the retail price.**

   a.   Markup percent of cost times cost equals the markup.

   b.   Divide the markup by the markup percent of cost.

   c.   Cost + Markup = Retail.

   Example: An item at a tack store had a markup of $60, which was 75 percent of the cost. Find the cost and the retail price.

$$0.75 \times \text{Cost} = \$60 \qquad\qquad \text{Cost} + \text{Markup} = \text{Retail}$$
$$\text{Cost} = \$60/0.75 \qquad\qquad \$80 + \$60 = \text{Retail}$$
$$\text{Cost} = \$80. \qquad\qquad \$140 = \text{Retail}.$$

6. **Given the cost and the markup percent of retail, find the retail or selling price.**

   a.   Retail = 100 percent. Subtract the markup percent from 100 percent to get the cost percent of retail.

   b.   Cost equals the cost percent of retail times retail price.

   Example: A retailer purchases steel posts from a manufacturer for $1.80 each. To get a fair profit, the retailer must get a profit of 40 percent of retail. Find the selling price needed to get the desired profit.

$$100\% - 40\% = 60\% \text{ (Cost percent of retail)}$$
$$\$1.80 = 0.60 \text{ (Retail)}$$
$$\$1.80/0.60 = \text{Retail}$$
$$\$3.00 = \text{Retail}.$$

7. **Given the desired retail price and the markup percent of retail, find the cost.**

   a.   Retail = 100 percent. Subtract the markup percent from 100 percent to find the cost percent of retail.

   b.   Cost equals the cost percent of retail times the retail price.

   Example: A retail buyer estimates that a particular tarp will sell for $35. A markup of 38 percent of retail is needed. What is the maximum amount the buyer can afford to pay for the tarp?

$$100\% - 38\% = 62\% \text{ (Cost percent of retail)}$$
$$\text{Cost} = 0.62 \times \$35$$
$$\text{Cost} = \$21.70.$$

8. **Given the amount of markup and the markup percent of retail, find the retail price and the cost.**

   a.   Markup percent of retail times retail equals markup.

b. Cost + Markup = Retail.

Example: On an item in a dairy supply store, the percent markup on retail is 37½ percent and the dollar markup is $15. Find the retail price and the cost.

$$0.375 \times \text{Retail} = \$15 \qquad \text{Cost} + \text{Markup} = \text{Retail}$$
$$\text{Retail} = \$15/0.375 \qquad \text{Cost} + \$15 = \$40$$
$$\text{Retail} = \$40. \qquad \text{Cost} = \$40 - \$15$$
$$\text{Cost} = \$25.$$

B. Cash Discounts

Credit ratings of businesses and individuals are influenced a great deal by the promptness of paying accounts as they come due. An incentive to pay promptly is the **cash discount**.

Manufacturers and distributors of goods allow a reduction in the net price of an invoice to induce buyers to make payments during a definite time period. Both the purchaser and the seller benefit; the purchaser pays a smaller amount and the seller is able to make use of the money sooner for buying more goods and paying manufacturing costs.

The most common way of expressing the payment terms of an invoice is called **ordinary dating**. The terms 3/10, n/30 mean that the purchaser may deduct 3 percent of the invoice price if payment is made within 10 days of the invoice date and that the full invoice price is due without discount in 30 days. A penalty may be charged if payment is not made within 30 days. More than one discount option may be offered, such as 3/10, 2/20, n/30. These terms mean that a 3 percent discount is allowed if payment is made in 10 days, 2 percent may be deducted if payment is made within 20 days, and the full amount is due in 30 days.

Two other forms of dating are:

1. **End of month (E.O.M.) dating.** If the terms 3/10, E.O.M. appear on an invoice dated August 16, the purchaser may qualify for a 3 percent discount if payment is made within 10 days after the end of August, or September 10.

   If an invoice is dated after the 25th of the month, the discount period is extended to 10 days after the end of the following month. On an invoice dated August 28 with terms of 3/10, E.O.M., the 3 percent discount will be allowed until October 10.

2. **Receipt of goods (R.O.G.) dating.** On an invoice having R.O.G. dating, the discount period is counted from the date the purchaser receives the goods rather than from the date of the invoice. An invoice dated November 22 and carries terms of 3/10, n/30, R.O.G. The purchaser received the goods on December 4. The purchaser may deduct 3 percent of the invoice price if payment is made by December 14 and the full invoice price, without discount, is due on January 3.

Exercise 4–2

MERCHANDISING AND RETAILING, CASH DISCOUNTS

Give answers to the nearest whole cent or to the nearest tenth of a percent.

1. Cost = $561.12; selling price = $801.60.

    a. Markup =   _____

    b. Markup percent on cost =   _____

    c. Markup percent on selling price =   _____

2. The markup percent on selling price is 72 percent. What is the percent markup on cost?

    _____

3. The markup percent on cost is 33⅓ percent. Find the percent markup on selling price.

    _____

4. Cost = $52.00; markup = 15 percent on cost. Find the selling price.

    _____

5. Selling price = $14.52; markup = 21 percent on cost. Find the cost.

    _____

6. Cost = $125.50; markup = 16⅔ percent based on selling price. Find the selling price.

_____

7. Selling price = $18.00; markup = 40 percent based on selling price. Find the cost.

_____

8. Selling price = $15; cost = $5. Find the markup based on cost.

_____

9. Selling price = $4.05; markup based on cost = 35 percent. Find the cost.

_____

10. An article costs $120 and sells for $160. Find the percent markup on cost.

_____

11. A roadside market operator purchases potatoes at 10.8 cents per pound from a grower. If a 28 percent markup on the selling price is needed to cover expenses, what price must be charged for the potatoes?

_____

12. Cash flow costs of raising soybeans are calculated by a farmer to be $8.11 per bushel. If the market price is currently $7.75 per bushel, what percent loss (based on cost) would the farmer take if the beans were sold now?

_____

_____

13. An implement dealer has a marked price of $46,251 on a combine. If the combine costs the dealer $30,834 and overhead expenses are $2,468, what is the percent of net profit on the selling price of the combine?

_____

14. The manager of a yard and garden center had been trying to sell, without success, a gas-powered edger/trimmer that cost $140 from the manufacturer. The marked price had a 60 percent markup on cost. By late fall, the machine had not been sold, was marked down 20 percent off the marked price and consequently sold. What percent profit was made on the cost of the edger/trimmer?

_____

15. A flower shop wants to sell potted geraniums for $1.77 each. If the shop needs to have a markup of 65 percent based on cost, what can the shop afford to pay for the geraniums?

_____

16. Custom rates of potato digging using a two-row potato harvester are $74.75 per hectare. If the operator estimates total costs of operation at $62.55 per hectare, what percent of profit over operating costs is made by the custom operator?

_____

17. The crimped oats sold by a horse-feed store has a 6¢-per-kilogram markup, which is 40 percent of the cost. What is the retail price of the oats?

_____

18. A livestock supply dealer purchased a 45-bushel hog feeder for $264 from the manufacturer. At what price should the feeder be marked so a 12 percent markdown can be allowed and still have a gross profit of 20 percent of the selling price?

_____

19. A bulk-fertilizer dealer placed the ad shown in Figure 4–3 in a local newspaper.

   a. What was the regular price of the auger?   _____

   b. Find the percent markdown on regular price.   _____

   c. If the regular price represents a 30 percent markup
      on the selling price, what was the cost to the dealer?   _____

   d. Find the percent of gross profit (sale price) on the cost.   _____

BUY NOW! SAVE $75

5-INCH TAILGATE AUGER

A real time saver at planting time

Gas-engine powered

Hopper fits 38-inch tailgate

Easy to install

SALE PRICE $478

BULK FERTILIZER COMPANY

Rural, U.S.A.

Figure 4–3

20. A milk processor purchases fresh whole milk from farmers and sells it to the supermarkets in retail cartons. The processor pays $14.33 per hundredweight for the milk and operating costs are $5.18 per hundredweight. What price per hundredweight must be charged the supermarkets if the processor wants a 15 percent net profit based on selling price?

_____

Solve the following problems. If results of calculations have more than two decimal places, round to two decimal places. Round percents to the nearest tenth of a percent.

Find the cash discount and net price paid on each of the following:

	Amount of Invoice	Terms	Date of Invoice	Date Paid	Cash Discount	Net Price
21.	$1,263.45	4/10, 2/30, n/60	Oct. 9	Oct. 15	_____	_____
22.	$ 645.30	3/10, n/30, E.O.M.	May 16	June 5	_____	_____
23.	$ 236.85	5/10, n/30, R.O.G. (goods received March 28)	March 19	April 5	_____	_____
24.	$ 931.06	4/10, 2/30, n/60	June 6	June 26	_____	_____
25.	$1,543.80	3/10, n/30, E.O.M.	July 24	Aug. 16	_____	_____

26. Members of a cooperative get a 2¢-per-gallon cash discount on fuel purchases if payment is made within 10 days of purchase. If a gallon of diesel fuel sells for $1.15, what is the rate of discount?

_____

27. A supplier of stock fence advertises 110-meter rolls of 1.2-meter-high fence, regularly priced at $85.60 per roll, at a discount price of $72.76. What rate of cash discount was received?

_____

28. An invoice in the amount of $1,500 for repair parts was received by a tractor repair shop. Terms of the invoice were 2/10, n/30. The shop owner qualified for the discount by borrowing the money, paying $6.53 of interest. How much was saved by borrowing the money to pay the invoice on time?

_____

29. A home and garden center purchased 15 lopping shears at $18.69 each from a supplier. The invoice has cash discount terms 3/10, n/30, E.O.M. and is dated March 16.

    a. What is the last date on which payment will qualify
       for the discount?   _____

    b. What is the amount that should be remitted if the
       discount is received?   _____

30. On March 30, a farmer ordered evergreen trees to plant in a farmstead windbreak. The farmer received an invoice from the wholesale supplier having a total price of $448.25 and having payment terms of 3/10, n/60, R.O.G. The shipment of trees arrived on April 21.

    a. What is the last day the discount can be received?   _____

    b. If payment is made on time to receive the discount,
       what is the net price for the trees?   _____

_____

A. Interest

Farmers, business people, and private citizens often borrow or lend money to finance purchases that could not readily be made if only cash were used. The rental charge for the use of money is called **interest**. **Simple interest** is usually used on short-term loans and **compound interest** used on long-term loans.

1. Simple interest

The credit instrument used in negotiating the borrowing or lending of money is the **promissory note**. It is an unconditional promise in writing by one person to another to pay a stated sum of money, usually plus interest, at a future date.

The **face** or **principal of the note** is the amount in dollars and cents for which the note is written. The **date of the note** is the day on which the note is written. The **maturity date** is the date on which the note is payable. The **time of the note** is the length of time in days, months, or years between the date of the note up to and including the maturity date. The **rate of interest** is the percent of the face or principal that is charged on an annual basis for the use of the money. **Maturity value** is the sum of the face value and the interest due on the note. **Simple interest** is calculated by multiplying the principal by the rate of interest and by the time of the credit instrument. In formula form we have:

I = principal × rate × time, or I = prt, where

    I = interest
    p = principal or face of the note
    r = rate of interest on an annual basis
    t = time stated in years or fraction of a year.

There are generally two types of simple interest, **ordinary** or **banker's interest** and **exact interest**. Ordinary interest uses 360 days as the denominator of the time fraction, while exact interest uses 365 days as the denominator of the time fraction. When time is less than a year, the numerator of the fraction is the number of days between the date of the note up to and including the maturity date. The denominator of the time fraction will be 360 or 365 depending on whether ordinary or exact interest is being calculated.

Example   What is the exact interest on a note that has a face value of $1,200 if the interest rate is 6 percent and the time is 65 days?

$$I = prt$$
$$I = \$1,200 \times \frac{6}{100} \times \frac{65}{365} = \$12.82.$$

Example: What is the ordinary interest on a note that has a face value of $1,200 if the interest rate is 6 percent and the time is 65 days?

$$I = prt$$
$$I = \$1,200 \times \frac{6}{100} \times \frac{65}{360} = \$13.00.$$

Example: Find the ordinary interest on a note dated June 20, which has a face value of $350 at 9 percent interest and is due on October 20. (June 20 to October 20 = 122 days.)

$$I = prt$$
$$I = \$350 \times \frac{9}{100} \times \frac{122}{360} = \$10.68.$$

2. Finding the principal, rate, or time if the interest is given

If we solve the interest formula, $I = prt$, for the other factors we get:

$$p = \frac{I}{rt}, \quad r = \frac{I}{pt}, \quad \text{and } t = \frac{I}{pr}.$$

Example: Find the principal necessary to produce $360 of ordinary interest in 180 days at 7½ percent interest.

$$p = \frac{I}{rt} \qquad p = \frac{\$9.30}{\$1,860 \times \frac{180}{360}} = \$9,600.00.$$

Example: Find the interest rate necessary to produce $9.30 of ordinary interest in 72 days if the principal is $1,860.

$$r = \frac{I}{pt} \qquad r = \frac{\$9.30}{\$1,860 \times \frac{72}{360}} = 0.025 \times 100 = 2.5\%$$

Example: Find the time necessary to produce $228.90 of ordinary interest at 4 percent from a principal of $16,350.

$$t = \frac{I}{pr} \qquad t = \frac{\$228.90}{0.04 \times \$16,350} = 0.35 \text{ year.}$$

3. Exact time

To calculate exact time, the exact number of days in each month must be calculated and then added together.

Example: Find the exact time for a loan dated April 16 and due on August 16.

14 days in April
31 days in May
30 days in June
31 days in July
<u>16 days</u> in August
122 days = exact time

Example: Find the due date (maturity date) of a 90-day loan (exact time) dated May 22.

9 days in May
30 days in June
<u>31 days</u> in July
70 days
Need <u>20 days</u> of August
90 days

Due date is August 20.

B. Commission

Commission is a salary payment made to a salesperson based on the amount of sales the salesperson makes for the employer. Commission may be earned under several different plans. **Straight commission** pays a stated rate of the sales regardless of the amount of sales. **Graduated commission** pays a higher rate for higher volumes of sales. **Salary and commission** is a plan wherein a salesperson receives a specific salary plus a percent of the sales made during the pay period.

Example: A salesperson is paid on a **straight commission** of 10 percent of the amount of sales. During a pay period, a salesperson sold $3,500 worth of merchandise. What is the amount of gross pay due the salesperson?

$3,500 × 10% commission = $350 gross pay.

Example: A field representative for an oil company is paid on a **graduated commission** of 10 percent of the first $5,000 of oil products sold, 12 percent of the next $7,500 sold, and 15 percent of sales above $12,500. What is the field representative's gross pay for a month in which $18,000 worth of products were sold?

10% × $5,000 = $ 500
12% × $7,500 = $ 900
15% × $5,500 = $ 825
$2,225 = gross pay.

Example: An implement dealer pays the sales personnel a **salary** of $250 per week **plus a commission** of 5 percent of the sales they make. What would be the gross pay of a salesperson selling $15,000 worth of machinery during a two-week period?

2 weeks at $250 per week = $ 500
5% of $15,000 sales = $ 750
$1,250 = gross pay.

C. Sales Taxes

Many of the states and some of the cities in the United States, levy taxes on retail sales of items such as tangible personal property, services, admissions, restaurant meals, public utilities, traveler's lodging, and printing. The tax is usually calculated as a percent of the purchase price and is collected by the seller.

For convenience in calculating the amount of sales tax due, the city or state agency will furnish the retailer with tax tables from which the salesperson can quickly determine the tax due. If not all items for sale by a retailer are taxable, the cash register is usually designed to record sales of taxable and nontaxable merchandise separately. In this case, the amount of taxable items plus the sales tax due on the items is recorded as gross sales of taxable items. Periodically, the sales tax on the gross sales is calculated and the sales tax sent to the proper agency.

Example: Gross sales of taxable items = $32,364.00 (includes 5 percent tax).

$32,364.00 is 105% of actual sales
$32,364.00 ÷ 1.05 = $30,822.86 (actual sales)

$32,364.00 gross sales
− $30,822.86 actual sales
$ 1,541.14 sales tax due.

D. Property Taxes

The funding for the support of public schools, fire departments, police, roads, parks, libraries, street lighting, etc., is derived primarily from property taxes, either on real property (land or anything attached to land) or personal property (inventories, equipment, stocks, bonds, etc.). Business enterprises are subject to taxes on both real property and personal property.

Finding the Tax Rate

The value of taxable property is usually determined by an employee of the taxing unit called a **tax assessor**. The value assigned to the property by the assessor is called the **assessed valuation**. The assessed valuation varies considerably from the fair market value of the property. The assessed valuation is usually a set percent of the fair market value.

To find the tax rate or tax levy for a taxing unit, the money or revenue needed to be raised (budget) is compared to the total assessed valuation of property in the taxing district. Expressed as a formula it is:

$$\text{Tax rate} = \frac{\text{Revenue to be raised}}{\text{Total assessed valuation}}.$$

The decimal quotient resulting from this calculation can be expressed in a number of ways: (a) cents per dollar — ¢/$, (b) dollars per $100 — $/$100, or (c) mills (a mill is 0.001 dollar).

Example: A town has a budget of $955,600 for the year and the assessed valuation of property in the city is $18,450,000. What tax rate must be set to raise this amount of revenue?

$$\text{Tax Rate} = \frac{\$955,600}{\$18,450,000} = 0.05179.$$

a. One cent = $0.01 $0.05179 \times \dfrac{1 \text{ cent}}{\$0.01}$ = 5.179 or 5.18¢/$.

b. $0.05179 \times \dfrac{\$100}{\$100}$ = 5.179 or 5.18/$100.

c. $0.05179 \times \dfrac{1 \text{ mill}}{\$0.001}$ = 51.79 or 51.8 mills.

To find the tax due on property having an assessed valuation of $8,000 in this taxing district using each of the types of tax rate, perform the following calculations.

a. 5.18¢/$1.00 × $8,000 = 41,400¢ $\times \dfrac{\$1.00}{100¢}$ = $414.40.

b. $5.18/$100 × $8,000 = $414.40.

c. 51.8 mills $\times \dfrac{\$0.001}{1 \text{ mill}}$ × $8,000 = $414.40.

To assure that there is no shortage of revenue, the last digit of each of the tax rates is rounded up even though the digit to the right of the digit to be rounded is less than five. For example: 5.173 cents would be rounded to 5.18 cents and 51.73 mills would be rounded to 51.8 mills.

Exercise 4–3

INTEREST, COMMISSION, TAXES

Round all dollar amounts to the nearest whole cent and percents to the nearest tenth of a percent.

1. Find the maturity date (date due) for a 90-day note dated May 21.

   _____

2. Find the exact number of days from July 16 to November 6.

   _____

3. Find the amount of ordinary interest due on a loan of $2,450 at 8½ percent for 90 days.

   _____

4. Find the amount of ordinary interest due on a loan of $4,160 at 9 percent from March 21 until July 6.

   _____

5. Find the exact interest on a loan of $2,095 at 10 percent interest dated November 21 and due on February 15.

   _____

_____

6. Find the principal of a loan if $369.24 of interest is earned in 270 days at 10 percent (ordinary interest).

_____

7. Find the rate of ordinary interest charged on a $14,500 note that earned $762.25 of interest in 236 days.

_____

8. Find the length of time (in days) needed for a loan of $2,368.42 at 9.5 percent (ordinary interest) to earn $150.

_____

9. A fertilizer bill in the amount of $1,845 was due on May 15, but the farmer was not able to make payment until July 6. If the dealer charges 12 percent interest on overdue bills, what amount was due on July 6 (ordinary interest)?

_____

10. The federal income tax return of a grain transport agency was audited and an accounting error was discovered, requiring the payment of $2,463 additional tax. If the Internal Revenue Service charged 9 percent (exact interest) for the period from April 15 (filing date) until July 30 (payment date), what was the total amount due?

_____

11. In order to finance the purchase of a $28,000 tractor, a farmer needs to borrow $13,480 to close the deal. If 9 percent interest is charged on a 90-day loan, what amount of ordinary interest would need to be paid?

_____

12. A "closing cost" fee is often charged for processing a loan. An apple grower borrowed $35,500 to build additional refrigerated storage facilities. If the lending agency charges legal fees of $200 plus ½ percent of the amount by which the loan exceeds $20,000, what is the amount of the "closing cost" on this loan?

_____

13. On June 8, a riding stable owner borrowed $10,840 to purchase additional riding horses. What rate of interest (ordinary) was paid if the loan plus interest was paid back on September 14 with a check for $11,188.20.

_____

14. A farmer in an area hit by a tornado received a disaster loan from the Small Business Administration in the amount of $340,000. Terms for the loan are 3 percent per annum (per year) on the first $250,000 and 7 percent on the remainder. Find the amount of ordinary interest due each year.

_____

15. Member of an implement dealer's sales force are paid a fixed salary of $500 per month, plus commission paid on the basis of the graduated commission system shown below.

   a. Find the total monthly earning of a salesperson selling
      $23,400 worth of equipment.   _____

   b. Find the total monthly earnings of a salesperson selling
      $40,930 worth of equipment.   _____

First $4,000	@	3%
$4,001 to $12,000	@	5%
$12,001 to $21,000	@	7%
$21,001 and over	@	8%

16. Prior to the spring planting season, a seed corn sales representative had sold 2,458 bushels of seed corn averaging $53.60 per bushel. If early sales get a 9.5 percent commission rate, what amount of commission was received?

_____

17. A field representative sells oil to farmers on a commission basis. If twelve 210-liter drums at $220.36 per drum and sixty-five 20-liter cans at $23.60 per can are sold and the representative gets a 12 percent commission, what is the amount of the commission received?

_____

18. A feeder pig sales barn charges a sales commission of $2.31 per head. If a farmer's pigs sold for an average of $38.50 per head, what rate of commission was charged?

_____

19. A landscape contractor purchased a 3.5-horsepower hole digger for $631.75, which included the 6 percent state sales tax. What was the price before tax?

_____

20. Sales tax amounting to $193.60 was charged on the purchase of $3,520 two-row forage harvester. What was the rate of sales tax?

_____

21. The cash price of a big round hay baler is $15,645. If a 5 percent state sales tax must be paid, what is the total cost of the baler?

_____

22. A drainage and tile contractor purchased a transit level for $1,005.94, which included a 5 percent state sales tax and a 1 percent municipal tax.

    a. What was the price of the level before tax?   _____

    b. What was the amount of the state sales tax?   _____

    c. What was the amount of the municipal tax?   _____

23. A garden center received a shipment of 75 canna bulbs costing $0.65 each, 150 tuberous begonia bulbs at $0.72 each, and 175 caladium bulbs at $0.42 each. If a 5 percent sales tax was charged on the order, what was the total amount due?

                                                          _____

24. For taxation purposes, a township assesses real estate at 35 percent of its market value. If a potato grower's land is assessed at $75,467, what is the market value of the land?

                                                          _____

25. Find the $/$100 rate of taxation of a school district having a budget of $896,680 in a county with $42,750,800 of assessed valuation. (Round up any part of a cent to the next larger cent.)

                                                          _____

26. Real estate taxes are based on levies made by the county, the township, and the local school district. If a turkey grower's real estate tax bill is $9,847.68 and 48 percent of the collected tax goes to the township and county, how much of the tax does the school district receive?

                                                          _____

27. A sheep rancher's personal property has an assessed valuation of $25,475. If the tax rate is 35.9 mills, what amount of tax is due?

_____

28. A truck-garden partnership has personal property with assessed valuation of $20,560. If the tax rate is $7.32 per $100 of valuation, what amount of tax is due?

_____

Exercise 4–4

APPLICATION EXTENSION

Calculating Germination Percentages

Unfortunately, seeds do not have 100 percent germination ability. Therefore, allow for this and plan over the calculated amount to provide the desired plant population. For example, if a sample of seeds has 90 percent germination, this means that 90 percent (or 90 out of 100) will germinate and grow to a mature plant. Therefore, if 100 plants are desired, plant 11 extra or 111 seeds to provide 100 plants.

Example: If a sweet corn population of 20,000 plants per acre is desired and the corn is to be planted in 30-inch rows, how far apart in inches will seeds need to be dropped? (Assume 90 percent field germination.)

Step 1. Calculate the total number of seeds that need to be planted per acre.

Number of desired plants ÷ germination rate = total seeds.

20,000 ÷ 90% or .9 = 22,222 seeds needed.

Step 2. Calculate linear inches of row per acre (see preceding section).

Square feet in 1 acre ÷ distance of rows in feet = linear feet.
Linear feet × number of inches in 1 foot = linear inches.

43,560 ÷ 2.5 = 17,424 linear feet.
17,424 × 12 = 209,088 linear inches.

Step 3. Calculate distance of seeds.

Linear inches ÷ total seeds = distance apart.

209,088 ÷ 22,222 = 9.41 inches apart.

Solve the following:

1. Sweet corn population of 22,000 plants per acre, 90 percent germination. How many actual seeds per acre are needed?

_____

2. If a sample of butterbean seeds have 85 percent germination, how many seeds should be planted to obtain a plant population of 30,000 plants per acre?

_____

3. Squash population of 10,000 seeds per acre, 80 percent germination, rows 36 inches apart. How far apart should seeds be dropped?

_____

4. How far apart should the seeds in question #1 be spaced if the rows are 36 inches apart?

_____

5. How far apart should the seeds in question #2 be spaced if the rows are 30 inches apart?

_____

6. One bag of watermelon seed contains 90,000 seeds. The desired plant population is 8,000 seeds per acre. With rows at 30 inches apart and the seed having 80 percent germination:

a. How far apart should the seeds be spaced?   _____

b. How many acres will one bag cover?   _____

5

INTERPRETATION AND ANALYSIS OF DATA

Key Terms

averages	extrapolation	median
bar graph	graphs	mode
circle graph	interpolation	standard deviation
continuous data	line graph	tables
discrete data	mean	

Career Connections

SOIL CONSERVATIONIST

A soil conservationist studies soil, water, and other natural resources. This person often works with landowners in using practices that conserve soil and water. A good knowledge of plants, especially row crops and pasture grasses, is needed.

Soil conservationists need a college degree. In college, they study agronomy or related areas. Soil conservationists should enjoy people and the outdoors. They usually work for state and federal government agencies that assist farmers and land developers.

The advancement of technology is accompanied by the accumulation and analysis of much data. There are a variety of ways to arrange data for study and interpretation. Sometimes it is desirable to find one or more numbers that are representative of the entire set of measures. These numbers are referred to as **averages**. The most frequently used averages are the **mean, median**, and **mode**.

The significance of averages of collected data depends on how well the average represents the whole set; that is, whether the numbers in the collected data vary greatly from the average or fall within a relatively narrow range from the average. The most frequently used measure of this variability is the **standard deviation**.

Tables having both a vertical and a horizontal reference are often used to present data. These tables are sometimes a summation of a large set of information. The given information can be extended by **interpolation** (for values between given values) and by **extrapolation** (for values beyond the given values).

If is often advantageous to use diagrams or scaled drawings to express data. These pictorial presentations are called **graphs**. The most commonly used graphs are the **bar graph**, the **line graph**, and the **circle graph**. One type of graph may give a more meaningful portrayal of certain collected data than another.

Discrete data, often referred to as exact data, allow for no interpretation between given values of data. Examples of discrete data are 6 truckloads, 470 bushels, 10 acres, and 17 kilograms. Discrete data can be counted. **Bar graphs** show this type of data well using either horizontal or vertical bars. If discrete data represent the various parts, which make up a total quantity, a **circle graph** is the best way to present the relationship.

Continuous data consist of measurements that could be made more precise or be taken at more frequent intervals. Examples of continuous data are temperatures taken over a period of time, gallons of fuel used, time spent on a project, etc. The best graph for presenting continuous data is the **line graph**.

PART I.   AVERAGES — MEAN, MEDIAN, MODE, AND STANDARD DEVIATION

A.   The **mean** (arithmetic mean) is the quotient obtained by dividing the sum of the measures by the number of measures in the data. To find the mean of a set of data:

Step 1.   Compute the sum of the data.

Step 2.   Divide the sum of the data by the number of measures being considered. (Note: If any measures occur more than once, the measures should be multiplied by the number of times they occur.)

Example: Find the mean of the data set: 25, 36, 27, 28, 31, 33

Step 1. $25 + 36 + 27 + 28 + 31 + 33 = 180$

Step 2. $180 \div 6 = 30$

We can think of the mean of a set of data as the "balancing point" much like a "see-saw" or "teeter-board."

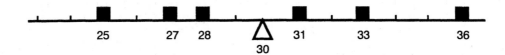

Example: Find the mean of the set of student test scores:

Number of students	Score
2	38
3	43
5	48
1	50

Step 1. $2 \times 38 + 3 \times 43 + 5 \times 48 + 1 \times 50 =$
          $76 \quad + \quad 129 \quad + \quad 240 \quad + \quad 50 \quad = 495$

Step 2. $495 \div 11 = 45$ points

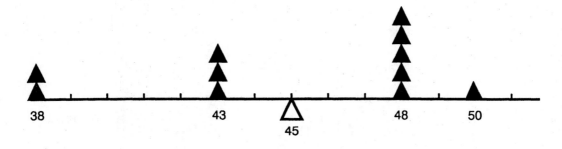

B. The **median** is the middle measure of a group of measures. To find the median of collected data:

Step 1.   Arrange the data in order of size, either from lowest to highest or vice versa.

Step 2.   Find the middle of the group of data.

    a.   If the number of measures is odd, the middle measure is the median.

    b.   If the number of measures is even, the median is the arithmetic mean of the two middle measures.

Example: Find the median of the data: 6, 9, 4, 5, 8, 5, 9, 3, 2, 7

Step 1.   2, 3, 4, 5, 5, 6, 7, 8, 9, 9

Step 2.   Number of data is even; median is arithmetic mean of 5 and 6.

$$\frac{5 + 6}{2} = 5.5 \quad \text{The median is 5.5}$$

Example: Find the median of the data: 93, 86, 72, 83, 92, 77, 85

Step 1.   72, 77, 83, 85, 86, 92, 93

Step 2.   Number of data is odd; median is the middle measure. Median is 85.

Example: Find the median of the following test scores on a 10-point quiz.

Score	1	2	3	4	5	6	7	8	9	10
Number of Students	0	1	0	1	2	4	3	4	4	3

Step 1.   2, 4, 5, 5, 6, 6, 6, 6, 7, 7, 7, 8, 8, 8, 8, 9, 9, 9, 9, 10, 10, 10

Step 2.   There are an even number of scores; median is arithmetic mean of the two middle measures. The median is 7.5.

C. The **mode** of a set of measures is the measure that appears most frequently in the data. To find the mode of a group of measures:

Step 1.   Determine the frequency of each measure in the set.

Step 2.   The mode is the measure that appears most frequently.

Example: Find the mode of the following data:

17, 13, 19, 13, 25, 19, 17, 19, 23, 15, 17, 13, 19

Step 1.   13: three times
            15: once
            17: three times
            19: four times
            23: once
            25: once

Step 2.   The measure 19 appears most frequently in the data. Nineteen
            is the mode.

D. When working with data, it is frequently important to know more than the measures of average. The **standard deviation** gives an idea of the dispersion of the set of values. If the standard deviation is small, the values are clustered relatively close to the mean. A large standard deviation indicates a more variable or widely dispersed set of data.

There are several similar formulas for computing the standard deviation. The formula presented here is the "shortcut formula" for computing the standard deviation for data obtained from samples. Since this formula does not use the mean in the computation of the standard deviation, it is equally easy to use whether the mean is a whole number or not.

$$\sqrt{\frac{\Sigma x^2 - \frac{(\Sigma x)^2}{n}}{n-1}}$$

Σ — indicates the operation of addition

x — represents the values of the data

x^2 — indicates that a number must be multiplied by itself (squared)

n — is the number of values

$\sqrt{}$ — indicates a square root that can be computed easily with the corresponding button on your calculator. (More will be explained about this in Chapter 6.)

Note that Σx^2 is not the same as $(\Sigma x)^2$. The notation Σx^2 means square the values first, then add; $(\Sigma x)^2$ means to add the values first, then square.

The following examples explain how to compute the standard deviation:

Two machine operators tilled the following numbers of acres during five sample time periods.

Operator A: 5, 6, 3, 6, and 5   Mean: 5.
Operator B: 8, 2, 3, 7, and 6   Mean: 5.1

Example: Find the standard deviation for the number of acres tilled by machinery operator A.

    Step 1.   Find the sum of the values.

$$5 + 6 + 6 + 3 + 5 = 25$$

    Step 2.   Square each value and find the sum.

$$25 + 36 + 36 + 9 + 25 = 131$$

    Step 3.   Substitute the numbers into the formula.

$$\sqrt{\frac{131 - \frac{(25)^2}{5}}{5 - 1}}$$

    Step 4.   Square the sum of the x's.

$$\sqrt{\frac{131 - \frac{625}{5}}{5 - 1}}$$

    Step 5.   Divide this product by n.

$$\sqrt{\frac{131 - 125}{5 - 1}}$$

    Step 6.   Subtract.

$$\sqrt{\frac{6}{4}}$$

    Step 7.   Divide by n – 1.

$$\sqrt{1.5}$$

    Step 8.   Find the square root with your calculator.

$$1.22$$

Example: Find the standard deviation for the number of acres tilled by machinery operator B.

    Step 1. $8 + 2 + 3 + 7 + 6 = 26$

    Step 2. $64 + 4 + 9 + 49 + 36 = 162$

    Step 3. $\sqrt{\dfrac{162 - \frac{26^2}{5}}{5 - 1}}$

Step 4. $\sqrt{\dfrac{162 - \dfrac{676}{5}}{5 - 1}}$

Step 5. $\sqrt{\dfrac{162 - 135.2}{5 - 1}}$

Step 6. $\sqrt{\dfrac{26.8}{4}}$

Step 7. $\sqrt{6.7}$

Step 8. 2.59

Since the standard deviation of the data from operator B is larger than the standard deviation of operator A, we know the data values for B are more "spread out." Also, as the standard deviation increases the mean becomes less useful as a representation of the data. Of the three "averages," the mean is influenced most by extreme high or low values.

Name_____ Date_____

Exercise 5–1

AVERAGES

Find the mean, median, mode, and standard deviation of the sets of data given in problems 1 through 4.

1. 34, 36, 41, 36, 33, 32, 29, 36, 35, 32, 34, 42

Mean	_____
Median	_____
Mode	_____
Standard Deviation	_____

2. 9, 3, 7, 14, 11, 9, 11, 7, 12, 9, 3, 9, 13, 5, 8

Mean	_____
Median	_____
Mode	_____
Standard Deviation	_____

3. Samples of barley delivered to a grain terminal had the following bushel test weights in pounds: 46, 50, 47, 49, 51, 50, 50, 51, 49, 48, and 48.

Mean	_____
Median	_____
Mode	_____
Standard Deviation	_____

4. A technician analyzing peanut samples at a crushing plant recorded the following percents of oil content in the kernel: 49.4, 44.8, 50.2, 48.7, 46.3, 44.8, 49.9, 43.4, 49.7, and 44.8.

Mean _____

Median _____

Mode _____

Standard Deviation _____

5. A student got scores of 45, 36, 54, 33, and 48 on the first five tests in math. What average score will be needed on the three remaining tests if the student needs an average of 45 points on all tests to get a passing grade in the course?

_____

6. A flower shop purchased the following plants for resale: six Swedish ivy at $4.25, seven philodendron at $2.75, five crotons at $3.80, eight hoya at $3.10, and four ferns at $6.70. What was the average price per plant?

_____

7. Five steers were sold with an average (mean) of 1,104 pounds each. If three of the steers weighed 1,216 pounds, 1,029 pounds, and 997 pounds, respectively, what is the average weight of the other two steers?

_____

8. A farmer sold 1,245 bushels of soybeans at $7.58 in May, 4,530 bushels at $7.31 in June, 3,785 bushels at $7.75 in July, and 2,495 bushels at $7.10 in August. What was the average (mean) price received per bushel?

_____

9. A landscape architect ordered rose bushes for a new rose bed. Included in the order were four bushes at $8.45 each, six bushes at $10.95 each, two bushes at $14.99 each, and a collection of nine bushes for $75.50. Find the average (mean) price per rose bush.

_____

10. If eight cattle are sold at auction for $.83/lb, $.60/lb, $.72/lb, $.39/lb, $.63/lb, $.75/lb, $.67/lb, and $.78/lb:

   a. What is the average price per pound?   _____

   b. Would you expect the standard deviation to be large or small? Why?   _____

   c. Compute the standard deviation.   _____

11. What is the average price per pound received for a 300-pound side of beef if 10 percent of it is sold for $2.39 per pound; 18 percent at $2.09; 24 percent at $1.95; 22 percent at $1.75; 14 percent at $1.45; 9 percent at $1.25; and 3 percent at $0.75?

_____

12. Workers at a fertilizer blending plant include 6 clerical workers, 30 production workers, and 4 salespersons. If the average weekly salary for all workers is $315, the average weekly clerical salary is $320, and the average production worker's salary is $290, what is the average weekly salary of the salespersons?

PART II. INTERPRETATION AND ANALYSIS OF TABLES

A. Tables

Tabular information generally is arranged so as to have two dimensions, one vertical and the other horizontal. In the example below, the time of day is presented vertically, with the corresponding temperatures given horizontally. Interpretation and analysis of the table involves the study of the vertical and horizontal information and establishing a relationship between the data given.

Example:

Temperatures recorded on August 17

6:00 a.m.	46°F
8:00 a.m.	52°F
10:00 a.m.	60°F
12:00 noon	70°F
2:00 p.m.	82°F
4:00 p.m.	87°F
6:00 p.m.	84°F
8:00 p.m.	77°F

Possible items of information that could be gleaned from the table are these:

a. Lowest temperature for the day was 46°F at 6:00 a.m.

b. The two-hour period having the greatest temperature increase was 12:00 noon – 2:00 p.m.

c. High temperature for the day was 87°F at 4:00 p.m.

d. The range of temperature for the day was 41°F.

Because tables are usually made up of information that is a small sampling of a larger set of data, we can extend the given information by the process of interpolation (for values between given data) and by extrapolation (for values beyond or outside the given data).

1. Interpolation

To interpolate between two items of data in a table:

Step 1.   Locate the given data that include the wanted information and list relevant information.

Step 2.   Find the difference between one of the given data and wanted information.

Step 3.   Find the difference between the given data located in Step 1.

Step 4.   Find the corresponding information in the other dimension of reference for the differences found in Steps 2 and 3.

Step 5.   Set up a proportion and solve.

Step 6.   Add or subtract the quantity found in Step 5 to or from the given data.

Example: Using the example from Part A, find the temperature at 9:30 a.m.

Step 1.   9:30 a.m. is between given data 8:00 a.m. and 10:00 a.m. The relevant data are:

Step 2.   9:30 a.m. – 8:00 a.m. = 1.5 hr

Step 3.   10:00 a.m. – 8:00 a.m. = 2 hr

Step 4.   Corresponding temperatures — 60° – 52° = 8°

Step 5. $\dfrac{1.5}{2} = \dfrac{x}{8°}, \dfrac{2x}{2} = \dfrac{\overset{4}{8} \times 1.5}{2}, x = 6°$

Step 6.   Because values are increasing, add 6° to 52° = 58°.

Temperature at 9:30 a.m. is 58°.

2. Extrapolation

Essentially the same process is used for extrapolation as is used for interpolation except that the wanted information is always at the "end" of the given data and the last two or first two sets of given data are used.

Example: Find the temperature at 8:30 p.m. in the example from Part A.

Step 1. 8:30 p.m. is located at the end of the table. The relevant data are:

```
                ┌─── 6:00 p.m. — 84° ───┐
         2 hr  ┤                        ├ 7°
2.5 hr ┤        └─── 8:00 p.m. — 77° ───┘          x
         │
         └────────── 8:30 p.m. —  ? ──────────┘
```

Step 2. 8:00 p.m. – 6:00 p.m. = 2 hr

Step 3. 8:30 p.m. – 6:00 p.m. = 2.5 hr

Step 4. $84° - 77° = 7°$

Step 5. $\dfrac{2}{2.5} = \dfrac{7°}{x}, \dfrac{2x}{2} = \dfrac{7 \times 2.5}{2}, x = 8.75°$

Step 6. Values of temperatures are decreasing at this point, so $84° - 8.75° = 75.25°$.

Temperature at 8:30 p.m. = 75.25°.

Exercise 5–2

TABLES

A. Interpretation of Tabular Data

Information relating to agriculture is presented in many tabular formats. In each of the following problems, use the tabular information to answer the questions accompanying the table.

1.

Soybeans: Area Planted by State and United States, 1997–99				
State	**Area Planted**			
	1997	**1998**	**1999**[1]	**1999/1998**
	1,000 Acres	*1,000 Acres*	*1,000 Acres*	*Percent*
AL	350	340	280	82
AR	3,650	3,550	3,500	99
DE	230	220	205	93
FL	47	35	30	86
GA	400	300	250	83
IL	10,000	10,700	10,800	101
IN	5,350	5,700	5,600	98
IA	10,500	10,500	10,900	104
KS	2,400	2,550	2,600	102
KY	1,240	1,220	1,170	96
LA	1,400	1,200	1,000	83
MD	530	470	470	100
MI	1,870	1,900	2,000	105
MN	6,600	6,900	6,900	100
MS	2,100	2,050	1,850	90
MO	4,900	5,100	5,100	100
NE	3,600	3,800	4,300	113

(Continued)

State	Area Planted			
	1997	**1998**	**1999**[1]	**1999/1998**
	1,000 Acres	*1,000 Acres*	*1,000 Acres*	*Percent*
NJ	133	115	110	96
NY[2]		100	110	110
NC	1,400	1,475	1,500	102
ND	1,150	1,550	1,400	90
OH	4,350	4,400	4,500	102
OK	340	470	470	100
PA	375	400	370	93
SC	580	540	480	89
SD	3,300	3,450	3,900	113
TN	1,240	1,250	1,250	110
TX	420	440	260	59
VA	510	500	500	100
WI	1,040	1,150	1,300	113
US	70,005	72,375	73,105	101

[1]Intended plantings in 1999 as indicated by reports from farmers.

[2]NY estimates began with 1998 crop year.

a. Which state had the most area planted in soybeans in 1998? _____

   Which state will plant the most in 1999? _____

b. How many acres are projected to be planted in New Jersey in 1999? _____

c. Which sate has the largest percent gain in 1999 compared to 1998? _____

d. How many states are planting fewer soybeans in 1999 than 1998? _____

e. Which state planted the same number of acres in 1997 and 1998? _____

Name_____ Date _____

2.

U.S. Agricultural Exports										
			Crops (crop year)				Livestock (calendar year)			
							Red Meat		Poultry	
Year	Corn	Wheat	Soybeans	Rice	Tobacco	Cotton	Beef	Pork	Broilers	Turkeys
	million bushels	million bushels	million bushels	million cwt	million pounds	(000) bales	million pounds	million pounds	million pounds	million pounds
1992	1,663	1,354	770	77	574	5,200	1,324	420	1,489	202
1993	1,328	1,228	589	75	458	6,860	1,275	446	1,966	244
1994	2,177	1,188	838	99	434	9,400	1,611	549	2,876	280
1995	2,228	1,241	851	83	462	7,680	1,821	787	3,894	348
1996	1,795	1,001	882	78	485	6,870	1,877	970	4,420	438
1997	1,504	1,040	870	85	484	7,500	2,136	1,044	4,664	598
1998	1,700	1,100	830	87	464	4,300	2,158	1,232	4,516	438
1999[1]							2,340	1,355	4,425	430

[1]Forecast. NASS, WAOB, & ERS (Information Hotline 1-800-727-9540)

a. Which livestock product had the most pounds exported in 1996? _____

b. Which livestock product had the greatest number of pounds exported for each year shown in this chart?   _____

   Which had the least?   _____

c. Can we compare quantity of various crop exports?   _____

   Why/Why not?

   Would any information allow us to compare crop exports?

d. Which year had the largest amount of corn exports?   _____

e. Has any crop consistently increased in exports?   _____

   Consistently decreased?   _____

3.

Vegetables: Acreage, Yield, Production, Price, and Value 1993–98, United States						
Crop and Year	**Acres**		**Yield per Acre**	**Total Production**	**Average Price**	**Total Value**
	Planted	**Harvested**				
				thousand	*dollars*	*thousand dollars*
Broccoli						
1993	119,400	119,200	101 cwt	12,059 cwt	25.80	310,675
1994	134,400	134,100	117 cwt	15,714 cwt	26.70	419,571
1995	129,600	129,400	122 cwt	15,815 cwt	28.00	443,304
1996	133,700	133,500	118 cwt	15,693 cwt	26.50	415,695
1997	130,800	130,800	129 cwt	16,880 cwt	28.50	481,459
1998	133,500	133,000	138 cwt	18,398 cwt	30.10	554,338

a. In which year were the most acres of broccoli planted? _____

b. How many acres of broccoli were harvested in 1994? _____

c. In which year were the same numbers of acres planted and harvested? _____

d. How could you explain fewer acres being harvested than were planted? _____

e. Which year had the highest yield per acre? _____

f. Which year had the highest average price? _____

Name_____ Date _____

4. Index numbers of prices received provides us with a method of comparing present prices with prices in previous years.

Index Numbers of Prices Received United States, May 1999, with Comparisons							
Index:	**1910–14 = 100**				**1990–92 = 100**		
		1998	**1999**		**1998**	**1999**	
Average:	**1990–92**	**May**	**April**	**May**	**May**	**April**	**May**
All Farm Products	639	656	610	630	103	96	99
All Crops	500	554	509	523	112	103	106
Food Grains	325	345	306	295	109	96	93
Feed Grains & Hay	363	385	328	333	108	92	93
Cotton	517	537	479	486	105	94	95
Oil-Bearing Crops	554	614	454	437	112	83	80
Fruit & Nuts	716	773	763	862	111	109	123
Com. Vegetables	698	841	856	836	126	128	125
Potatoes & Dry Beans	540	555	520	552	110	103	109
Other Crops	493	532	532	532	108	108	108
Livestock & Products	768	734	690	716	96	90	93
Meat Animals	1021	888	822	844	87	81	83
Dairy Products	799	817	774	793	102	96	99
Poultry & Eggs	282	304	292	309	108	104	110

Examples: Cotton prices in May 1999 are 486% of the average price from 1910–14, but are only 95% of the price from 1990–92.

Poultry and egg prices in 1990–92 are 282% of the average price of 1910–14. Their price in April 1999 is 292% of the price in 1910–14 and 104% of the price of 1990–92.

a.   "All farm products" in May 1999 were _____ compared to 1910–14

and had a value of _____% of 1990–92.

b.   Did any item decrease in price between 1910–14 and 1999? _____

c.   Did any item decrease in price between 1990–92 and 1999? _____

d.   Which crop increased by the greatest percent between 1910–14 and May 1999?   _____

e.   Which was the only livestock product to increase in price between 1990–92 and May 1999?   _____

5.   True–False   (Mark with T or F).

Field Tests—Characteristics of Soybean Varieties				
Variety	Height (inches)	Weight of 100 Seeds (grams)	Protein (percent)	Oil (percent)
A	43	19.4	41.0	18.7
B	39	19.6	42.6	18.0
C	36	17.5	40.7	19.6
D	41	17.0	41.5	18.5

a.   The variety having the highest oil content also has the largest seeds.   _____

b.   The variety having the largest seeds has the lowest protein content.   _____

Name_____ Date _____

c. The shortest variety also has the lowest oil content.   _____

d. The average height of all varieties is 39.25 inches.   _____

e. The range of protein content for the four varieties is 1.9 percent.   _____

B. Interpolation and Extrapolation

Using the data given in each of the following problems, interpolate or extrapolate to find the answers to the accompanying questions.

1. Interpolate or extrapolate to find the equivalent temperatures.

Equivalent Fahrenheit and Celsius Temperatures	
F°	C°
104°	40°
68°	20°
32°	0°
–4°	–20°

a. 45°F = _____ °C

b. –15°C = _____ °F

c. 120°F = _____ °C

d. –40°C = _____ °F

e. 75°F = _____ °C

2. A garden center has the following schedule for sale prices of quantity purchases of tulip bulbs. The owner wants to extend the quantity options to 36 bulbs, 60 bulbs, 72 bulbs, and 120 bulbs. What price should each of these quantities have?

Quantity	Price
12 bulbs	$ 4.19
24 bulbs	7.49
48 bulbs	13.59
96 bulbs	24.49

36 bulbs — $_____

60 bulbs — $_____

72 bulbs — $_____

120 bulbs — $_____

3. The table below gives the kernel spacings for various planting rates for corn planted in rows 24 inches apart.

Planting Rate	Kernel Spacing
17,800	14.7 inches
20,000	13.0 inches
22,200	11.4 inches
26,700	9.7 inches
31,100	8.3 inches

a. What kernel spacing is needed for a planting rate of 18,400?_____

b. If the kernel spacing were 7.4 inches, what is the corresponding planting rate? _____

c. Find the planting rate for a kernel spacing of 10.2 inches. _____

d. What kernel spacing would give a planting rate of 16,900? _____

148

Name_____   Date_____

4. The table below gives the number of trees in a hectare of peach orchard with various area requirements per tree.

Peach Orchard Population	
Area Needed per Tree	**Trees per Hectare**
30.3 m²	330 trees/ha
37.2 m²	267 trees/ha
46.3 m²	216 trees/ha
57.8 m²	173 trees/ha

a. What area is allowed per tree if there are 235 trees per hectare? _____

b. If an orchard is to be planted giving each tree 34.8 square meters of area, what number of trees can be planted per hectare? _____

c. If the area requirement per tree is 62.4 square meters, find the tree population per hectare. _____

d. If there are only 154 trees per hectare, how many square meters of area are allowed per tree? _____

A. Bar Graph

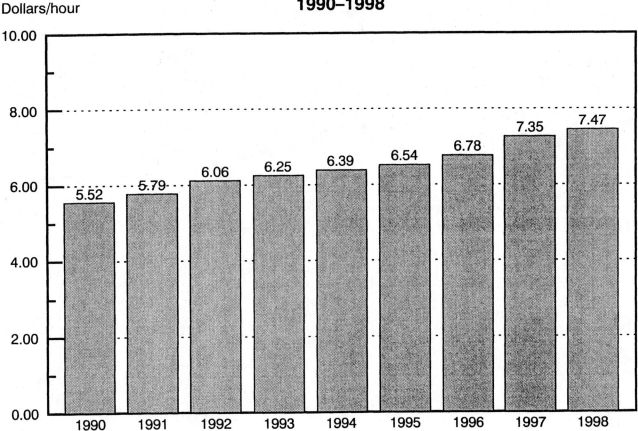

**Hired Farm Workers: Average U.S. Wage Rates
1990–1998**

Dollars/hour

NASS, Livestock & Economics Branch, (202) 720-3570

Some Observations:

• The average wage of hired farm workers has increased each year from 1990–98.

• In 1990, the average wage was $5.52 per hour and in 1998 the average wage was $7.47 per hour.

• It would be possible to compute the amount or percent increase in average annual wage between any two years from 1990 to 1998.

B. Line Graph

Number of Farms and Average Farm Size 1975–1998
United States

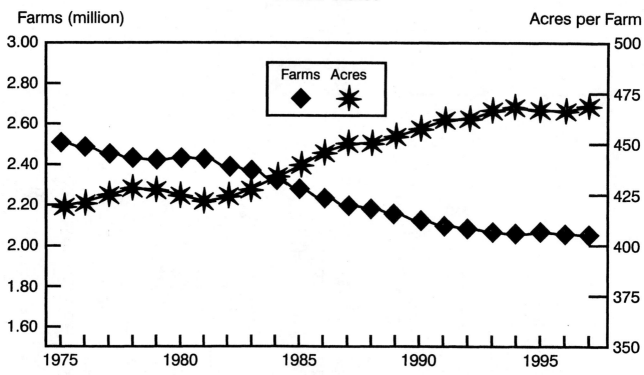

1975–96 estimates are for June 1 date. 1997 estimates are for the entire year.

NASS, Livestock & Economics Branch, (202) 720-3570

Some Observations:

• The number of farms has decreased between 1975 and 1998 while the acres per farm has increased.

• There were approximately 2,500,000 farms in 1975 and slightly over 2,000,000 farms in 1998.

• Farm size increased from less than 425 acres per farm to almost 475 acres per farm.

C. Circle Graph

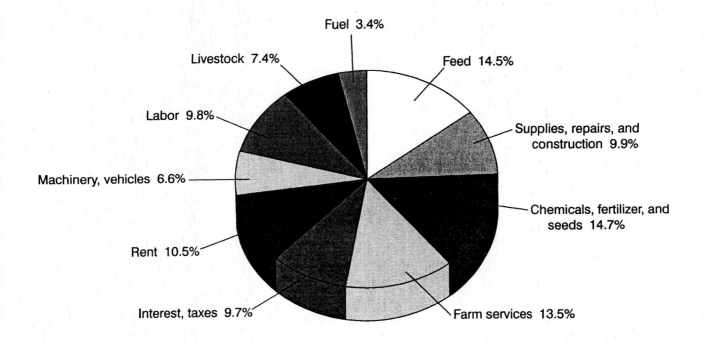

Some Observations:

- Expenditures for feed are almost the amount spent for chemicals, fertilizers, and seeds.

- Interest and taxes are 9.7% of expenditures.

- Almost twice as much is spent for chemicals, fertilizers, and seeds as is spent for livestock.

Name_____ Date _____

Exercise 5–3

INTERPRETATION OF GRAPHED DATA

1. The bar graph below shows a comparison of the average composition of colostrum, whole milk, and reconstituted milk replacer when used as calf feeds. Study the graph and answer the accompanying questions.

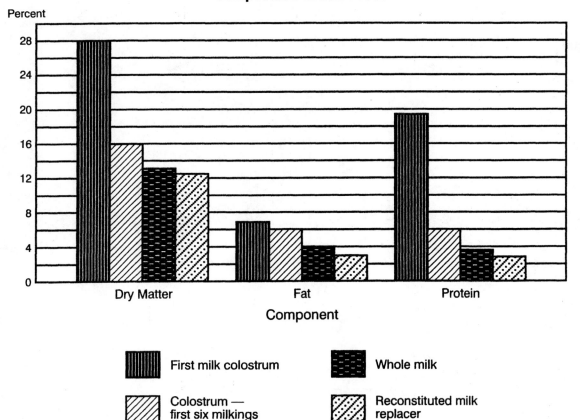

Composition of Calf Feeds

a. The calf feeds vary least from each other in their _____ content.

b. _____ has about 1.75 times as much dry matter as _____ .

c. The calf feed that is superior in all components is _____ .

d. If a calf is fed 2 pounds of first milk colostrum, how much protein does the calf get?
   _____ .

155

2. In a country located near an urban area, farm real estate values have changed as shown in the following bar graph.

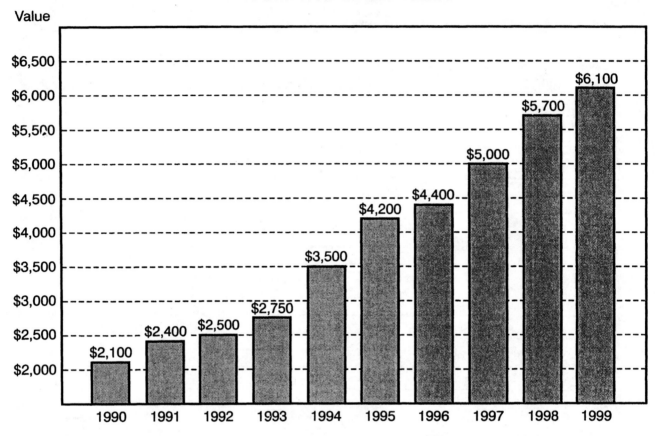

Farm Real Estate Values

a. What was the percent of value increase during the decade 1990–1999? _____

b. 1995 land values are what percent of 1990 values? _____

c. The land value change was the same dollar amount between years 1994 and 1995 and between years 1997 and 1998. What is the percent of increase between 1994 and 1995? _____

What is the percent of increase between 1997 and 1998? _____

Explain why the percent of increases are different.

Name_____ Date_____

3. The crude protein content of the major high-protein feed supplements is compared in the bar graph below. Use the data given to answer the accompanying questions.

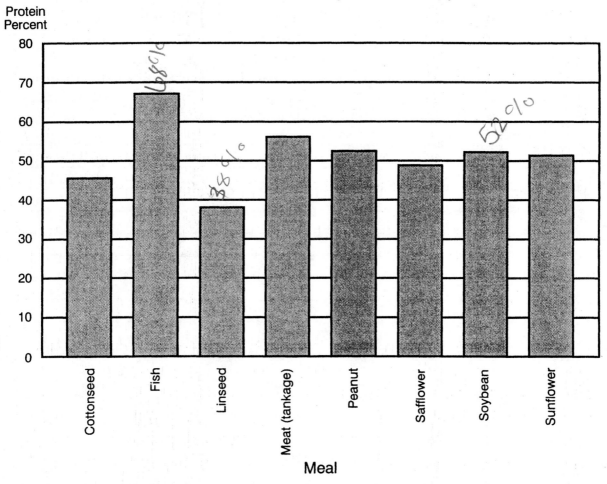

Crude Protein Content of High-Protein Supplements

a. The meal having the highest percent of crude protein is _____.

b. If soybean meal costs $220.40 per ton and linseed meal costs $155.00 per ton, which meal is the more economical source of protein? _____

c. The meal having approximately two-thirds the crude protein content of fish meal is

_____.

d. The meal of plant origin having the lowest crude protein content is _____.

_____

4.

Recorded Temperatures on December 8

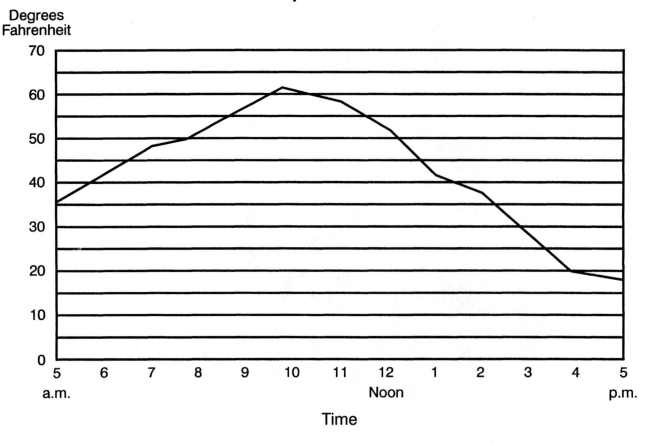

Degrees
Fahrenheit

Time

a. The lowest temperature for the 12-hour period is _____.

b. The high temperature for the 12-hour period is _____.

c. The range of temperatures for the day is _____ degrees.

d. The high temperature for the day occurred at (time) _____.

e. The temperature at 5:00 P.M. was _____ than the temperature at 5:00 A.M.

Name_____   Date _____

5.

How the Food Dollar Is Spent

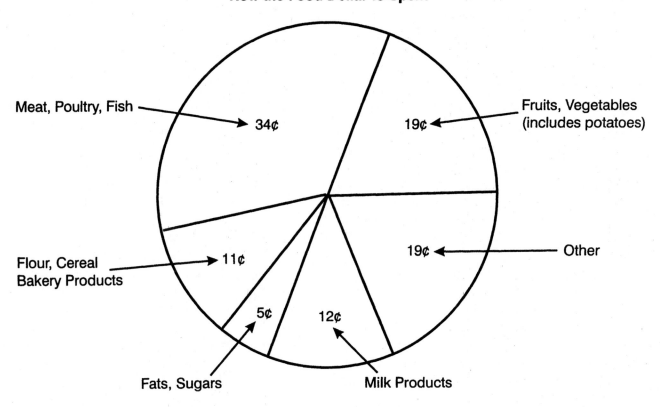

Meat, Poultry, Fish — 34¢

Fruits, Vegetables (includes potatoes) — 19¢

Flour, Cereal Bakery Products — 11¢

Other — 19¢

Fats, Sugars — 5¢

Milk Products — 12¢

a. The largest portion of the food dollar is spent for

_____.

b. Expenditures for meat, poultry, and fish are about _____ times the amount spent for milk products.

c. _____ percent of the food dollar is spent for fats and sugars.

d. _____ percent of the food dollar is spent for meat, poultry, fish, and milk products.

e. _____ percent as much is spent for fruit and vegetables as is spent for meat, poultry, and fish.

6

INTRODUCTION TO ALGEBRA

Key Terms

absolute value
algebraic expressions
binomials
constant
decimal-number
exponents
integers
monomial
negative numbers
number line
numerical coefficient

polynomials
positive numbers
rational numbers
scientific notation
similar terms
square root
symbols of grouping
term
trinomials
variable

Career Connections

ENTREPRENEURSHIP

An entrepreneur is a businessperson who creates new opportunities. Often, an entrepreneur develops a new, innovative product or service and enters business for himself or herself. Some people buy a franchise for an agribusiness or become a distributor/dealer for agricultural products. Entrepreneurs must understand management and marketing. They must be willing to take a risk. Consider starting your own business—you will be amazed at what you learn!

PART I. OPERATIONS WITH INTEGERS

A. Definitions

1. **The number line.** On the **number line** below, numbers located to the right of a given number are greater than (>) the given number; numbers located to the left of a given number are less than (<) the given number.

2. **Integers.** The counting numbers or whole numbers of arithmetic and their opposites or negatives are called **integers**. Zero is also considered an integer. These numbers are also a part of the set of numbers called rational numbers.

3. **Rational numbers** take their name from "ratio." They are fractions (both positive and negative) and any decimal that can be written as a fraction.

4. **Absolute value.** The **absolute value** of a number may be thought of as the distance a number lies from 0 on the number line, irrespective of direction. We use the symbol | | to indicate absolute value: for example, $|+4| = |-4| = 4$.

5. **Positive and negative numbers.** Numbers greater than zero are prefixed with a plus sign (+) and are called **positive**. Numbers less than zero have a minus sign (–) and are said to be **"negative."**

B. Addition and Subtraction of Signed Numbers

1. **Numerals having the same sign.**

    Step 1.   Find the sum of the absolute values of the numbers.

    Step 2.   Prefix the sum with the common sign of the addends.

    Example: Compute the sum of –2 and –5.

    The absolute value of –2 is 2; absolute value of –5 is 5.

    Sum of the absolute values: $2 + 5 = 7$

    Common sign of the addends is minus (–).

    Sum of –2 and –5 = –7

Example: Compute the sum of +6 and +5.

$$|+6| + |+5| = 6 + 5 = 11$$

$$(+6) + (+5) = +11$$

2. Numerals having different signs.

Step 1.   Find the difference of the absolute values of the numbers.

Step 2.   Prefix the difference with the sign of the addend having the greater absolute value.

Example: Add +8 and –5.

Absolute value of +8 is 8; absolute value of –5 is 5. Difference of absolute values: $8 - 5 = 3$

The sign of the addend having the greater absolute value is +, so $(+8) + (-5) = +3$.

3. Finding the difference of two integers.

Add the negative of the subtrahend (number to be subtracted) to the minuend.

Example: Subtract: $(-4) - (-5)$
The negative of –5 is +5.
$(-4) - (-5) = (-4) + (+5) = +1$

Example: Subtract $(-4) - (+5)$
The negative of +5 is –5.
$(-4) - (+5) = (-4) + (-5) = -9$

C. Multiplication and Division of Signed Numbers

1. To compute the product of two or more signed numbers:

a.   Find the products of the absolute values of the factors.

b.   Prefix the product with a **plus sign** if the number of negative factors is **even** (none, two, four, etc.) and with a **minus sign** if the number of negative factors is **odd** (one, three, five, etc.).

Example: $(+2)(-4)(-6)(+3) = +144$

Example: $(+3)(-3)(+5) = -45$

---

2. **To compute the quotient of two signed numbers:**

    a.   Find the quotient for the absolute values of the two numbers.

    b.   Prefix the quotient with a **plus sign** if the two numbers are **both positive or both negative** and with a **minus sign** if **one is positive and the other is negative.**

        Example: $\dfrac{+6}{-3} = -2$

        Example: $\dfrac{-8}{-4} = +2$

D. Signs for Fractions

If a fraction is considered as possessing three signs, that is, the sign of the numerator, the sign of the denominator, and the sign of the complete fraction, then any two of the signs can be reversed without changing the value of the fraction.

Example: $\quad +\dfrac{(+2)}{(+3)} \;=\; +\dfrac{(-2)}{(-3)} \;=\; -\dfrac{(+2)}{(-3)} \;=\; -\dfrac{(-2)}{(+3)}$

Exercise 6–1

OPERATIONS WITH INTEGERS

1. At an appropriate point on the number line below, locate the following numbers:

    (a) $-\dfrac{5}{2}$ (b) $+3$ (c) $1\dfrac{1}{8}$ (d) $-4\dfrac{3}{4}$ (e) -0.5 (f) $+2.7$

Insert the appropriate symbol >, <, or = between the following pairs of numbers:

2. $+3$ _____ -6 3. $+43$ _____ $+34$ 4. $\dfrac{3}{8}$ _____ 0.375

5. 0 _____ -2 6. $-\dfrac{1}{4}$ _____ $+\dfrac{1}{8}$ 7. $+0.16\dfrac{2}{3}$ _____ $-\dfrac{1}{6}$

8. $+2$ _____ $-\dfrac{5}{2}$ 9. -6.3 _____ -4.2

Write the equivalent arithmetic numbers for the following absolute values:

10. $|-0.5| = $ _____   11. $|+16| = $ _____

12. $\left|+\dfrac{1}{4}\right| = $ _____   13. $|-19| = $ _____

Add:

14. $\begin{array}{r} +\dfrac{5}{6} \\ -\dfrac{3}{8} \\ \hline \end{array}$ 15. $\begin{array}{r} +21.10 \\ -36.76 \\ \hline \end{array}$ 16. $\begin{array}{r} -\dfrac{84}{15} \\ +\dfrac{58}{30} \\ \hline \end{array}$

17. −1,732
    −2,045

18. +18.21
    − 2.38

19. +46
    −28
    +93

20. + 1.3
    − 4.8
    −10.4
    +11.9

21. $-\dfrac{3}{4}$
    $+\dfrac{5}{6}$
    $-\dfrac{7}{8}$

22. −16
    −21
    −35
    +98
    −13

23. +16.01
    −44.65
    −20.99
    +81.37

Subtract the lower number from the upper number:

24. −116
    +201

25. +176
    − 89

26. −16.76
    −21.43

27. +83
    +79

28. −19.38
    +21.01

29. $-39\dfrac{4}{5}$
    $+21\dfrac{3}{4}$

30. $+\dfrac{17}{3}$
    $-\dfrac{21}{16}$

31. $-15\dfrac{3}{8}$
    $-12\dfrac{5}{7}$

32. +6,754
    +4,867

33. −12.55
    −64.38

Multiply:

34. $(-2)(-25)(-7) =$

35. $(+46.34)(-60.02) =$

36. $(-2)(-6)(+3)(-9)(-4) =$

37. $\left(+\dfrac{2}{7}\right)\left(-\dfrac{1}{3}\right)\left(+\dfrac{7}{8}\right)\left(-\dfrac{3}{2}\right) =$

38. $(-3.71)(-4.28) =$

39. $(-1.06)(+0.036) =$

Divide:

40. $(-48) \div (-16) =$

41. $(+6,784) \div (-41) =$

42. $\left(-\dfrac{5}{6}\right) \div \left(+\dfrac{7}{12}\right) =$

43. $(+8.32) \div (-0.32) =$

44. $(-21,684) \div (-7.8) =$

45. $\left(-2\dfrac{2}{3}\right) \div \left(-4\dfrac{2}{3}\right) =$

Perform the indicated operations:

46. $\dfrac{(-4) + (-8)}{(-6)(-4)} =$

47. $\dfrac{16(-3) + 6}{(-2) - (+2)} =$

48. $\dfrac{(-16)(+4) \div (-3)}{(-7)(-2)} =$

49. $\dfrac{(+21) \div (-7)}{(-81) \div (+27)} =$

Indicate the proper sign to make the pairs of fractions equivalent:

50. $(+) \dfrac{-4}{+15} = (\ \ ) \dfrac{4}{-15}$ 51. $(-) \dfrac{-8}{+13} = (\ \ ) \dfrac{-8}{-13}$ 52. $(+) \dfrac{+2}{-3} = (\ \ ) \dfrac{-2}{-3}$

53. The recorded daily high and low Fahrenheit temperatures during a week in December are given below. Find the number of degrees difference between each day's high and low temperature.

	High	Low	Difference
Sunday	+45	+27	_____
Monday	+29	+16	_____
Tuesday	+20	–2	_____
Wednesday	0	–22	_____
Thursday	–8	–24	_____
Friday	+10	–8	_____
Saturday	+25	–4	_____

54. The table below gives grain futures for sunflowers. A **positive Change** quotation indicates that today's closing price is higher than yesterday's closing price. A **negative Change** quotation indicates that today's closing is lower than yesterday's closing.

a. What was the closing price of July futures yesterday? _____

b. What was the closing price of November futures yesterday? _____

c. What is the difference in price of the closing July and January quotation? _____

d. The greatest difference between July and November futures occurred in which quotation (Open, High, Low, or Close)? _____

	Open	High	Low	Close	Change
July	12.35	12.58	12.25	12.58	+0.26
November	12.57	12.70	12.48	12.48	–0.11
January	12.83	12.92	12.75	12.90	+0.17

PART II. OPERATIONS WITH ALGEBRAIC EXPRESSIONS, POWERS, SYMBOLS OF GROUPING

A. Definitions

1. **Constant.** A numeral is called a **constant** because it represents a definite number or quantity.

2. **Variable.** Letters of the alphabet (a, b, c, etc.) are used in algebra to represent an unknown number. Because the letter could represent any number, it is called a **variable**.

3. **Algebraic expressions.** The expression, $5ay - 4x + 3$, is an **algebraic expression** consisting of three **terms** connected by signs of addition and subtraction. A single term is often called a **monomial**.

4. **Numerical coefficient.** The number part of an algebraic term is called a **numerical coefficient**.

5. **Similar terms.** Terms are said to be alike or **similar** when they have exactly the same literal factors (letter parts).

6. **Polynomials.** Any algebraic expression having one or more terms is called a **polynomial**. Polynomials consisting of two algebraic terms are called **binomials**, and polynomials having three terms are called **trinomials**.

B. Computation of the Sum and Difference of a Polynomial

1. **Finding the sum and difference of like terms.**

   Step 1.   Group like terms.

   Step 2.   Add or subtract the numerical coefficients of the terms.

   Step 3.   Affix to the numerical coefficient the common literal parts of the term.

   Example: Add $4x + 2y + 6$.
   There are no like terms so the expression remains as is.

   Example: Add $3m + 4n + 2 - 6n$.
   $3m + 4n - 6n + 2 = 3m - 2n + 2$

2. Finding the sum of several polynomials.

Step 1.   Arrange addends so like terms are in the same column.

Step 2.   Add each column of like terms.

Example: Add $(-4a + 2b - 3c) + (a - 6b + 5c) + (7a + b + c)$.

$$
\begin{array}{r}
-4a + 2b - 3c \\
a - 6b + 5c \\
\underline{7a + b + c} \\
4a - 3b + 3c
\end{array}
$$

3. Finding the difference of two polynomials.

Step 1.   Arrange the minuend and the subtrahend so like terms are in the same column.

Step 2.   Indicate the subtraction operation by reversing the sign of the subtrahend.

Step 3.   Add the like terms.

Example: $(7x + 6y) - (5x + 2y)$

$$
\begin{array}{r}
7x + 6y \\
^{(-)}_{+} 5x \quad ^{(-)}_{+} 2y \\
\hline
2x + 4y
\end{array}
$$

C. Finding the Product of Polynomials

1. Finding the product of powers.

Repeated multiplication of a number or variable by itself is often expressed by means of an **exponent**. $2 \cdot 2 \cdot 2 \cdot 2$ can be written 2^4. The 4 is the exponent and 2 is the base. 2^4 is usually read "two to the fourth power." The exponent tells how many times to use the base as a factor in the multiplication. If c^4 means $c \cdot c \cdot c \cdot c$ and c^2 means $c \cdot c$, then $c^4 \cdot c^2 = c \cdot c \cdot c \cdot c \cdot c \cdot c$ or c^6. **We can find the product of powers by adding the exponents.**

Step 1.   Add the exponents of the factors.

Step 2.   The product is written as the common base with an exponent equal to the sum of the exponents.

Example: $x^3 \cdot x^6 = x^9$

---

2. Multiplying several monomials.

Step 1.   Find the product of the numerical coefficients.

Step 2.   Find the product of the literal parts by adding the exponents of the variables.

Example: $(6a^2b^2c)(-2abc^3)(-3a^3b)$

$(6)(-2)(-3)(a^2 \cdot a \cdot a^3)(b^2 \cdot b \cdot B)(c \cdot c^3)$

$36a^{2+1+3}b^{2+1+1}c^{1+3} = 36a^6b^4c^4$

3. Multiplying a polynomial by a monomial.

Step 1.   Multiply each term of the polynomial by the monomial.

Step 2.   Add the products.

Example: $2a^3b(5ab^3 - 3a^4b)$

$(2a^3b)(5ab^3) = 10a^4b^4$

$(2a^3b)(-3a^4b) = -6a^7b^2$

$10a^4b^4 - 6a^7b^2$

4. Multiplying a polynomial by a polynomial.

Step 1.   Write one polynomial directly above the other.

Step 2.   Multiply each term of the top polynomial by each of the terms in the bottom polynomial, arranging like terms in the same column.

Step 3.   Add the like terms to complete the product.

Example: Multiply $(3a - 4c)$ by $(2a + 6c)$.

$$\begin{array}{r} 3a - 4c \\ 2a + 6c \\ \hline 6a^2 - 8ac \\ + 18ac - 24c^2 \\ \hline 6a^2 + 10ac - 24c^2 \end{array}$$

171

Example: Multiply $(2a + 3b + 8c)$ by $(4b - 3c)$.

$$
\begin{array}{r}
2a + 3b + 8c \\
4b - 3c \\
\hline
8ab + 12b^2 + 32bc \\
- 9bc - 6ac - 24c^2 \\
\hline
8ab + 12b^2 + 23bc - 6ac - 24c^2
\end{array}
$$

5. Multiplying binomial by binomial (special case).

In algebra, we frequently multiply a binomial times a binomial. There is a shortcut for this procedure and it can be remembered by the letters FOIL.

Step 1.   Multiply the **first** terms of each binomial.

Step 2.   Multiply the terms on the **outer** sides of the parentheses.

Step 3.   Multiply the terms to the **inner** sides of the parentheses.

Step 4.   Multiply the **last (or second)** terms of each binomial.

Step 5.   Combine like terms. They will usually be the terms found in steps 2 and 3.

Example: $(x + 3)(x - 2)$
$X \cdot X - 2 \cdot X + 3 \cdot X + 3 \cdot (-2)$
$x^2 - 2x + 3x - 6$
$x^2 + x - 6$

Example: $(x - 6)(x - 3)$
$x^2 - 3x - 6x + 18$
$x^2 - 9x + 18$

D. Symbols of Grouping

Parenthesis ( ), brackets [ ], and braces { } are frequently used in grouping terms in algebra to show the order in which the mathematical operations are to be performed. Positive or negative signs, which precede a *symbol of grouping*, may be considered either as affecting each term as it is removed from within the symbols of grouping or as affecting the result of the indicated operations within the symbols of grouping.

1. Removing symbols of grouping from an algebraic expression.

a.   A symbol of grouping preceded by a plus sign may be removed without changing the signs of the terms enclosed.

b.   When removing a symbol of grouping preceded by a minus sign, change the sign of EVERY term enclosed by the grouping symbols.

Example: $4p - [-q + (p - 3q) + 4q] - p$

$4p - [-q + p - 3q + 4q] - p$

$4p - [-4q + 4q + p] - p$

$4p - (p) - p$

$4p - p - p = 4p - 2p = 2p$

2.   Inserting symbols of grouping in algebraic expressions.

This process is the reverse of the removal process.

a.   Symbols of grouping that are to be preceded by a plus sign may be inserted into the expression without changing the signs of the terms enclosed.

b.   Algebraic terms that are to be enclosed by a symbol of grouping preceded by a minus sign must have their signs reversed if they are to remain equivalent to the original expression.

Examples: $-x + 2y = +(-x + 2y)$

$-x + 2y = -(x + 2y)$

$x - 2y = -(-x + 2y) = -(2y - x)$

E. Finding the Quotient of Polynomials

1. Finding the quotient of powers.

Step 1.   Subtract the exponent of the divisor from the exponent of the dividend.

Step 2.   Write the quotient as the common base with an exponent equal to the difference of the exponents.

Example: Divide x^5 by x^8, $x \neq 0$.

$$\frac{x^5}{x^8} = x^{5-8} = x^{-3} \text{ or } \frac{1}{x^3}$$

Note: a negative exponent in the numerator can be expressed as a positive exponent in the denominator.

Example: Divide a^9 by a^5, $a \neq 0$.

$$\frac{a^9}{a^5} = a^{9-5} = a^4$$

Example: Divide b^4 by b^4.

$$\frac{b^4}{b^4} = b^{4-4} = b^0 = 1$$

Note: Any power having a non-zero base and an exponent of 0 is equal to 1.

2. Finding the quotient of two monomials.

Step 1. Express the dividend and the divisor as a fraction with the dividend as the numerator and the divisor as the denominator.

Step 2. Find the quotient of the numerical coefficients.

Step 3. Find the quotient of the like literal parts by subtracting the exponent of the divisor from the exponent of the dividend.

Example: Divide $32a^3b^4c$ by $8ab^2$.

$$\frac{32a^3b^4c}{8ab^2} = 4a^{3-1}b^{4-2}c^{1-0}$$

Quotient $= 4a^2b^2c$

Example: Divide $-9a^2b^2c^3$ by $3ab^4c$.

$$\frac{-9a^2b^2c^3}{3ab^4c} = -3a^{2-1}b^{2-4}c^{3-1}$$

Quotient $= -3ab^{-2}c^2$ or $-\dfrac{3ac^2}{b^2}$

3. Finding the quotient of a polynomial divided by a monomial.

Step 1. Divide each term of the polynomial by the monomial.

Step 2. Add the quotients of each term for the total quotient.

Example: Divide $14a^2bc^3 - 7ab^4c$ by $7abc$.

$$\frac{14a^2bc^3}{7abc} - \frac{7ab^4c}{7abc}$$

$2a^{2-1}b^{1-1}c^{3-1} - 1a^{1-1}b^{4-1}c^{1-1}$

$2ac^2 - b^3$

Example: Divide $21x^4y - 15x^2y^3$ by $3x^3y^2$.

$$\frac{21x^4y}{3x^3y^2} - \frac{15x^2y^3}{3x^3y^2}$$

$7x^{4-3}y^{1-2} - 5x^{2-3}y^{3-2}$

$7xy^{-1} - 5x^{-1}y$ or $\dfrac{7x}{y} - \dfrac{5y}{x}$

F. Exponents and Symbols of Grouping

Signs of grouping, such as parenthesis, are often used to indicate the sequence of mathematical operations with exponents.

In the expression, $(-x)^2$, note that the exponent is located outside the parenthesis and that the mathematical operation will be $(-x)(-x) = +x^2$.

In the expression, $(-x^2)$ or $-x^2$, note that the exponent is located inside the parenthesis and that the mathematical operation will be $-x \cdot x$ or $-x^2$.

Example: $3^2 + (-2)^2 = 3 \times 3 + (-2)(-2) = 9 + 4 = 13$

Example: $2^3 - (-2^2) - 3^2 = 2 \times 2 \times 2 - (-2 \times 2) - 3 \times 3$

$\qquad = 8 - (-4) - 9 = 8 + 4 - 9 = 3$

Exercise 6–2

OPERATIONS WITH ALGEBRAIC EXPRESSIONS, POWERS, SYMBOLS OF GROUPING

Collect like terms and combine by adding or subtracting coefficients:

1. $3r + 2s - 5t + t - 2s + 2t - 5s + 3r + s$ _____

2. $a - 4c + 5b + 2c - 3a + 7b + 4c + 5a$ _____

3. $mn + 3mp - 5np + 3mn - 2np - 7np + 3mn + 4np$ _____

Rearrange the following expressions in column form and add:

4. $(7r - 4s + 3t), (2r - 8s + 9t), (-5r - 9s - 6t)$ _____

5. $(7a - 5b + 8c), (2a - 7b - 5c), (-9a + b - 4c)$ _____

Subtract:

6. $(x - 3y) - (2x + 5y)$ _____

7. $(-5r^2 - 7st) - (3st + 2t^2)$ _____

Subtract the second expression from the first:

8. $(-5a + 4b - 7c)$, $(4a + 3b + 2c)$ _____

9. $(6x - 3y + 2z)$, $(3x + 2y - z)$ _____

10. $(6m - 2n + 5p)$, $(3m + 5p - r)$ _____

Multiply:

11. $x^6 \cdot x^3$ _____

12. $2a^3 \cdot 3a \cdot 4a^3$ _____

13. $(-2x^3)(4x^2)(3x^4)$ _____

14. $(4m^2n)(-7m^2n^4p)(2m^3p^5)$ _____

15. $(-3rs)(2r^2s + 5st^2 - 2s^2t)$ _____

Multiply and simplify:

16. $3x - 8$
    $5x + 9$

17. $2b - 4x$
    $5b + 3x$

18. $6x^2 + 4xy - 5y^2$
    $2x - 3y$

19. $(x + 5)(x + 6)$

20. $(2x + 3)(x - 1)$

21. $(x - 7)(2x + 3)$

Remove grouping symbols and simplify:

22. $7x + \{6 - [y + 4 - 3(x + 3) - 2y] + 2\}$ _____

23. $a + 3b - [3b - a - (b - 3a) + b]$ _____

24. $(r - 3) - [s + 6 - (2r - 5s + 10)] + 3$ _____

25. $4m - \{n - [p - (n + p) + 2n] - 3m\} + p$ _____

26. $(3x + 3y) - [(9y - 4x) - 7x + 6y)]$ _____

Write the equivalent of the following expressions, enclosing the last three terms in parenthesis preceded by a minus sign:

27. $3x + 4y - 6r + 9$ _____

28. $7a - b + 4c + 8$ _____

Write the equivalent of the following expressions, enclosing the last two terms in parenthesis preceded by a plus sign:

29. $2mn + 6n + 7m^2n + 7$ _____

30. $3ax + 4by - 5ry + 1$ _____

Divide:

31. $y^7 \div y^4$ _____   32. $a^6 \div a^8$ _____

33. $2^4 \div 2^2$ _____   34. $b^7 \div b^7$ _____

35. $\dfrac{32p^5q^2}{4p^4q}$ _____   36. $\dfrac{39m^4n^8}{13m^6n^2}$ _____

37. $\dfrac{4x^2y^4}{8xy^3}$ _____   38. $\dfrac{63a^3b^2}{9a^2b^2}$ _____

39. $(27c^3d^2 + 18c^6d - 36cd^4) \div (9c^2d^2)$ _____

40. $\left(n^2 - \dfrac{1}{2}mn - \dfrac{3}{4}m^5n^2\right) \div \left(\dfrac{1}{4}n\right)$ _____

41. $\dfrac{24x^2y^2 - 18x^4y + 4x^3y^3}{6x^2y^2}$ _____

42. $\dfrac{2.8a^2b^2 + 3.5a^3b^5 - 4.9a^2b^6}{-7a^2b^4}$ _____

43. $\dfrac{-16m^4n^2 - 4m^3n + 40m^6n^5}{-0.4m^3n}$ _____

Perform the indicated operations and simplify:

44. $\dfrac{-4(-3)(-2) + (3)^2}{(-2^3) - (-3)}$ _____

45. $\dfrac{(-5)^2 + (-7^2)}{(6)(-2^2)(-2)}$ _____

46. $\dfrac{(-4^2) + (-3)(-2)(10) - (-3)}{2(13 - 7)^2 - (-5)}$ _____

47. $(-3)^3 - 2(-4 - 2)^2 + (-2)^2$ _____

PART III. SQUARE ROOT, SCIENTIFIC NOTATION, EVALUATION OF ALGEBRAIC EXPRESSIONS

A. Square Root

1. When we look for the **square root** of a number, we are looking for a number that when squared (multiplied by itself) will give us back the original number. The symbol for indicating square root is the radical sign, $\sqrt{}$.

    Example: $2^2 = 2 \times 2 = 4$, so $\sqrt{4} = \sqrt{2^2} = 2$

    $(-2)^2 = (-2)(-2) = +4$. Therefore, $\sqrt{4} =$ either $+2$ or -2. Most frequently we are looking only for the positive square root of a number and $\sqrt{9}$ usually means $+3$.

    Example: $s^2 = 16$ means $s \times s = 16$

    Therefore, $\sqrt{s^2} = \sqrt{16}$ or $s = \sqrt{16}$, $s = 4$.

    Example: $\sqrt{\dfrac{16}{25}} = \dfrac{\sqrt{16}}{\sqrt{25}} = \dfrac{4}{5}$

    Example: $(\sqrt{16})^2 = 4^2 = 16$

    When the number under the radical is a number whose square root is not readily recognizable, most all calculators will compute the value.

2. Square root is used in the solution of problems involving some familiar formulas.

    Area of a square: $A = s^2$, $s = \sqrt{A}$ (Formula 1)

    Area of a circle: $A = \pi r^2$, $r = \sqrt{\dfrac{A}{\pi}}$ (Formula 2; use 3.14 for π)

    Volume of a cylinder: $V = \pi r^2 h$, $r = \sqrt{\dfrac{V}{\pi h}}$ (Formula 2)

    The Pythagorean Theorem states that the square of the hypothenuse of a right triangle is equal to the sum of the squares of the other two sides.

    $c^2 = a^2 + b^2$, $c = \sqrt{a^2 + b^2}$ (Formula 4)

    $a^2 = c^2 - b^2$, $a = \sqrt{c^2 - b^2}$ (Formula 5)

    $b^2 = c^2 - a^2$, $b = \sqrt{c^2 - a^2}$ (Formula 6)

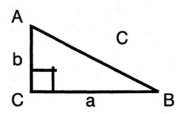

B. Scientific Notation

1. A very useful application of the exponent is in the writing of very large and very small numbers. To simplify the reading, writing, and computation of these numbers, we use scientific notation. **To write a number in *scientific notation*, we first move the decimal point from its place or indicated place (after the last digit) in ordinary notation to a place to the right of the first non-zero digit.** This is called **standard position**. We then count the number of places the decimal point was moved and use that number as a power of ten. If the decimal point was moved **to the left** to get it into standard position, **the exponent is positive**; if it was moved **to the right** to get it into standard position, **the exponent is negative**. For example: The number 123,000 is written 1.23×10^5 and the number 0.00286 is written 286×10^{-3}. The first part of the number in standard position is called the **decimal-number** part.

2. To add or subtract numbers in scientific notation, the power of ten of each of the numbers must be the same. If they are not the same, the power of ten must be changed so that they are alike.

   Example:　Add 1.7×10^3 meters and 1.693×10^5 meters expressing sum as 10^3. First change 1.693×10^5 to 169.3×10^3 meters, then add:

   $$\begin{array}{r} 1.7 \times 10^3 \text{ meters} \\ +\quad 169.3 \times 10^3 \text{ meters} \\ \hline 171.0 \times 10^3 \text{ meters} \end{array}$$

3. To multiply numbers written in scientific notation, we multiply the decimal-number parts together and add the exponents of the powers of ten.

   Example:　Multiply 255,000,000 by 0.0000156.

   First change each number to scientific notation:

   $2.55 \times 10^8 \times 1.56 \times 10^{-5}$

   $2.55 \times 1.56 = 3.98$ (rounded to three digits)

   $10^8 \times 10^{-5} = 10^{8-5} = 10^3$

   Our product is 3.98×10^3.

4. To divide numbers written in scientific notation, we divide the decimal-number parts and subtract the exponents of the powers of ten of the divisor from the exponents of the powers of ten of the dividend. As in the other arithmetical operations, the answer is usually rewritten into standard form if it is not already in that form.

   Example:　Divide 42,400 by 0.00656.

First change each number to scientific notation:

$4.24 \times 10^4 \div 6.56 \times 10^{-3}$

$4.24 \div 6.56 = 0.646$ (rounded to three digits)

$10^4 \div 10^{-3} = 10^{4-(-3)} = 10^{4+3} = 10^7$

The quotient is 0.646×10^7 or in standard form, 6.46×10^6.

Note that $10^0 = 1$.

C. Evaluation of Algebraic Expressions

When the values of the variables in an algebraic expression are known, the value of the entire expression can be found by substituting the known values in the expression and performing the indicated mathematical operations.

Example: Find the value of the expression $\dfrac{ab}{\sqrt{d}} - \dfrac{6bc}{a^2}$ if $a = -8$, $b = \dfrac{1}{4}$, $c = -32$, and $d = 16$.

Substituting, we have:

$$\frac{(-8)\left(\dfrac{1}{4}\right)}{\sqrt{16}} - \frac{6\left(\dfrac{1}{4}\right)(-32)}{(-8)^2} = \frac{-2}{4} - \frac{-48}{+64} = -\frac{1}{2} - \left(-\frac{3}{4}\right)$$

$$= -\frac{1}{2} + \frac{3}{4} = +\frac{1}{4}$$

Exercise 6–3

SQUARE ROOT, SCIENTIFIC NOTATION, EVALUATION OF ALGEBRAIC EXPRESSIONS

Find the square root:

1. $\sqrt{121} =$

2. $\sqrt{169} =$

3. $\dfrac{\sqrt{36}}{\sqrt{81}} =$

4. $(\sqrt{144})^2 =$

5. $\sqrt{100 + 125} =$

6. $\sqrt{256} =$

Use your calculator to find the square root of the following numbers. Round the root to two decimal places.

7. $\sqrt{97.48}$

8. $\sqrt{134.6}$

9. $\sqrt{0.876}$

10. $\sqrt{983.5}$

Solve the following problems using the square root formulas given in the introduction to square root.

11. How high on the side of a barn will a 15-foot ladder reach if the foot of the ladder is placed 9 feet away from the barn wall? Use formula (F6).

_____

12. A circular riding arena located near a stable contains 11,456 square feet of space. How far is it directly across the center of the arena? Use formula (F2).

_____

_____

13. The sliding door of a machine storage building is to be braced by joining opposite corners of the door by means of a metal strip. If the door measures 9 feet by 15 feet, what length of metal strip is needed? Use formula (F4).

_____

14. A square research plot for garden vegetables has an area of 212 square meters. What is the perimeter of (distance around) the plot? Use formula (F1).

_____

15. How long a rafter is required for a building 36 feet wide if the rise is 12 feet and there is no overhang? Use formula (F4).

_____

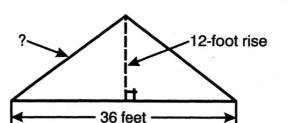

16. What diameter would a steel grain bin 20 feet high need to have to hold 6,500 bushels of corn? One bushel of corn occupies 1.25 cubic feet of space. Use formula (F3).

_____

Add:

17. 3.64×10^2 pounds + 1.18×10^1 pounds = _____ $\times 10^2$ pounds.

_____

186

Name_____ Date _____

Subtract:

18. 4.30×10^4 centimeters $- 8.59 \times 10^3$ centimeters = _____ $\times 10^4$ centimeters.

Perform the indicated operations. (Give the result in scientific notation (standard position), rounding the decimal number part to three digits.

19. $0.186 \times 5{,}280 \times 12$ _____

20. $7.46 \times 1.10 \times 10^{-3} \times 4.91 \times 10^2$ _____

21. $53.4 \times 10^4 \div 0.0502$ _____

22. $65{,}900 \div 8.03 \times 10^{-3}$ _____

23. $0.00291 \times 0.00798 \times 0.0000321$ _____

24. $\dfrac{7.45 \times 10^6 \times 8.21 \times 10^{-3}}{0.000862 \times 47{,}290}$ _____

25. $\dfrac{4.69 \times 10^{-4} \times 1.846 \times 10^3 \times 2.71 \times 10^2}{653 \times 0.000236}$ _____

26. $\dfrac{0.0000675 \times 34.8}{9.1 \times 10^1 \times 5.836 \times 10^3}$ _____

27. A dam built for generating electrical power and for supplying irrigation water contains about 1.12×10^7 cubic yards of concrete. If concrete weighs about 4.9×10^3 pounds per cubic yard, how many tons does the dam weigh?

_____

28. If the greatest rate of plant growth is about 3.0×10^{-3} centimeters per second, what is the shortest time in hours that a plant can grow 12.6 centimeters?

_____

29. An agricultural survey indicates there are 10,850,000 dairy cows in the United States. If each dairy cow drinks an average of 13 gallons of water daily, how many gallons of water are required annually by the dairy cows in the United States?

_____

30. Corn production in the United States is currently at the 7.082×10^9 bushel level annually. If this amount of corn is grown on 6.997×10^7 acres, what is the average yield of corn per acre?

_____

Evaluate each of the algebraic expressions below if a = 5, b = –3, c = 2, d = –4, e = –6, f = ½, g = –¼, h = 2¹/₃, and i = 0.

31. $a - b + c - 3d$

_____

32. $\dfrac{ghi + 2abc}{ab}$

_____

33. $b^2 + c^2 + 2bc + \sqrt{7}(a + c)$

_____

34. $3f + 2g - 5h$

_____

35. $5bd + 6fg$

_____

36. $\dfrac{c(a + 2d^2)}{\sqrt{d^2}}$

_____

37. $5f^3 - 4f^2g + 3fg^2$

_____

38. $2a + \dfrac{7b}{h} - \dfrac{3e + 5d}{c}$

_____

39. $\dfrac{bh - ei}{h^2}$

_____

40. $\dfrac{4ab}{ce}$

_____

_____

7

LINEAR EQUATIONS

Key Terms

balance scale
elimination by addition
elimination by substitution
equation

inverse operations
Pearson Square method
solving equations
system of equations

Career Connections

WATER TESTING

Water from wells, rivers, and lakes frequently must be tested. All water from the earth contains some substances. Testing shows if the substances are harmful to people and animals. This allows a person to know if using the water is a hazard.

The location of water-testing facilities varies. Testing may be done on site at a stream, lake, or other source. Sometimes, water samples are collected and tested in a laboratory. Since the reliability of a test can be no better than the sample tested, a good sample is needed. Poor sample quality can provide results that are misleading.

Anyone specializing in this field must have a good knowledge of mathematics, chemistry, and biology.

A. Variables and Equations

In solving many types of problems it is convenient to use a letter to stand for the number whose value is to be found. This letter is often called a **variable** because it could stand for any number. After introducing a letter to represent the unknown number, a statement of equality may be written in which the letter is used just as the number it represents would be used if it were known. This statement of equality is called an **equation**. Each of the numbers or algebraic expressions in an equation is called a **term** of the equation. Terms are separated from each other in the equation by plus or minus signs. For example: $7x = 6 + 2x$.

B. Solving Equations

The process of finding the value of the unknown or variable in an equation is known as **solving the equation**.

Think of an equation as a **balance scale**.

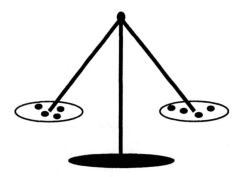

It is balanced when we begin and the balance must be maintained. Any operation performed on one side must be completed on both sides.

1. An equation is solved with a series of **inverse operations**. (Addition and subtraction are inverses. Multiplication and division are inverse operations.)

Examples: $x + 7 = 5$
         $\underline{-7}\quad \underline{-7}$ (subtract 7)

         $-x = -2$

         $X - 3 = 9$
         $\underline{+3}\quad \underline{+3}$ (add 3)

         $X = 12$

190

Example: Since 2x implies 2 times x, 2x = 6 would be solved by dividing by 2.

$$\frac{2x}{2} = \frac{6}{2}$$

$$x = 3$$

Example: Since $\frac{x}{4}$ implies x divided by 4,

$$\frac{x}{4} = 5$$

Is solved by multiplying by 4.

$$4 \cdot \frac{x}{4} = 5 \cdot 4$$

$$x = 20$$

2. Before beginning to solve with inverse operations, both sides of an equation must be simplified. This is accomplished by moving parentheses and/or combining like terms.

Example: $3(2x + 5) - 2 = 10 + 3$

$6x + 15 - 2 = 10 + 3$ Multiply to remove parentheses.

$6x + 13 = 13$ Combine like terms.
$\underline{-13 \quad -13}$ Subtract 13.

$6x = 0$

$$\frac{6x}{6} = \frac{0}{6}$$ Divide by 6.

$x = 0$ Zero is a real number and can be an answer.

3. If there are variables and constants on each side, the variables should be "moved" to one side and the constants "moved" to the other side by adding or subtracting.

Example: $5x - 5 = 2x + 4$
$\underline{-2x \qquad -2x}$ Subtract 2x from both sides.

$3x - 5 = 4$
$\underline{+5 \quad +5}$ Add 5 to both sides.

$$3x = 9$$

$$\frac{3x}{3} = \frac{9}{3} \qquad \text{Divide by 3.}$$

$$x = 3$$

4. In any equation with fractions, multiply every term by the lowest common denominator to cancel all denominators.

Example: $\quad \dfrac{x}{2} - 4 = \dfrac{x}{3}$

$$\overset{3}{\cancel{6}}\frac{x}{\cancel{2}} - \overset{6}{\cancel{6}}4 = \frac{x\cancel{6}^{2}}{\cancel{3}} \qquad \text{Multiply by LCD :6}$$

$$
\begin{array}{rcl}
3x - 24 & = & 2x \\
\underline{-2x} & & \underline{-2x} \qquad \text{Subtract 2x.} \\
x - 24 & = & 0 \\
\underline{+24} & & \underline{+24} \qquad \text{Add 24.} \\
x & = & 24
\end{array}
$$

C. Steps for Solving Equations

Step 1.   If there are fractions, multiply all terms by LCD.

Step 2.   Simplify both sides of equation.

Step 3.   "Move" variables to one side of equation.

Step 4.   "Move" constants to other side of equation.

Step 5.   Divide if necessary to obtain one of the variables.

Exercise 7–1

SOLVING EQUATIONS

Solve for indicated variable and check by substitution.

1. $6(x + 2) = 20$

2. $3x + 5 = 9x + 8$

3. $12a + 3 = -2a + 17$

4. $4(b + 2) = 24$

5. $7(c - 1) + 4 = 11$

6. $4y - 2 = 3y + 4$

7. $3(2y - 5) = 2(4y - 3) - 21$

8. $2(5m + 8) = 5(3m + 7) + 1$

9. $0.6a - 1.6 = 0.2a + 0.4$

10. $\frac{3}{5}a = 33$

11. $\frac{2y}{3} = 3.9$

12. $5 = b + 3\frac{1}{4}$

13. $\frac{1}{2}r + 3 = 31$

14. $7x - 1.8 = 1.7$

15. $4(y - 1) + 5(y + 1) = 3(y + 4) + 1$

16. $5.17y - 4.89 = 1.4174$

17. $3(x - 1) + x = 2(x - 1) - 3$

18. $6a - (a - 7) = 22$

19. $2(3p - 6) + 4 = 22 - 2(p - 1)$

20. $\frac{4y}{5} - 2 = 10$

21. $\dfrac{3b}{4} - \dfrac{2}{3} = \dfrac{5}{12}$

22. $\dfrac{a}{2} + \dfrac{3a + 1}{5} = \dfrac{a + 3}{10}$

23. $(x - 1) - (x - 4) = -(2x - 5)$

24. $\dfrac{4x + 3}{3} - \dfrac{2x + 3}{12} = 2$

25. $\dfrac{1}{3}(n + 4) - \dfrac{2}{5}(n - 8) = -3$

26. $\dfrac{s}{3} - \dfrac{7}{6} + \dfrac{2s}{5} = \dfrac{3s}{4}$

27. $(3a - 8) - (a - 4) = 0$

28. $x - 4 = \dfrac{1}{5}x$

29. $2(x - 5) - 3(2x + 3) = -23$

30. $\dfrac{1}{3}x - 2.4 = 3.6$

PART II. SOLVING STATED PROBLEMS

The ultimate goal of studying basic algebra is to enable us to analyze, set up, and solve stated problems that involve an unknown. The general steps needed to solve stated problems are:

Step 1.   Determine from the problem what we wish to know. A variable is assigned to that unknown.

Step 2.   Identify the relationship between the given information and the unknown. Eliminate all extra material.

Step 3.   If possible, draw a chart, diagram, or picture of the problem.

Step 4.   Develop an algebraic equation relating the known information and the unknown quantity.

Step 5.   Solve the equation for the variable or unknown.

Step 6.   Verify the solution by substituting the value back into the original problem.

The following problems are examples of the common types of stated problems:

A. Mixture Problems

Mixture problems involve the blending or mixing of two or more kinds of objects or materials to get some desired combination.

Example: A mixed bird feed sells for $1.00 per pound at a pet supply center. The bird feed consists of a commercial bird seed that sells at 95 cents per pound, millet seed that sells at 95 cents per pound, and organic sunflower seed that sells at $1.20 per pound. The mixture contains twice as many pounds of sunflower seeds as millet and the remainder is made up of the commercial bird seed. How many pounds of each type of seed should be mixed for a 10-pound bag?

Cost Amount						
	$\begin{array}{c}0.95\\ x\end{array}$	+	$\begin{array}{c}1.20\\ 2x\end{array}$	+	$\begin{array}{c}0.95\\ 10 - (x + 2x)\end{array}$	=	$\begin{array}{c}1.00\\ 10\end{array}$

$$10 = \text{total weight of mixture}$$
$$x = \text{the \# of millet seed}$$
$$2x = \text{the \# of sunflower seed}$$
$$10 - 3x = \text{the \# of commercial bird seed}$$

Equation: $0.95x + 1.20(2x) + 0.95(10 - 3x) = 1.00(10)$

$$0.95x + 2.40x + 9.5 - 2.85x = 10$$

$$.5x = 10 - 9.5$$

$$.5x = .5$$

$$x = 1\# \text{ millet seed}$$

$$2x = 2\# \text{ sunflower seed}$$

$$10 - 3x = 10 - 3$$

$$= 7\# \text{ commercial bird seed}$$

Example: How many pounds of ground ear corn testing 9.3 percent protein (dry matter basis) should be mixed with 100 pounds of a 48.7 percent protein supplement to make a dairy feed having a protein content of 18 percent?

The equation will be made up of terms each of which represents an amount of protein.

Amount of protein in ear corn + amount of protein in the supplement = amount of protein in the total weight of dairy feed mixed.

Percent Amount	9.3% x	+	48.7% 100	=	18% x +100

$$100 = \text{pounds of protein supplement}$$
$$x = \text{number of pounds of ear corn}$$
$$100 + x = \text{total weight of mixed dairy feed}$$

$$0.093x + 0.487(100) = 0.18(100 + x)$$

$$0.093x + 48.7 = 18 + 0.18x$$

$$48.7 - 18 = 0.18x - 0.093x$$

$$30.7 = 0.087x$$

$$352.87\# = x$$

Example: A farmer has $25,500 invested in two blended stock and bond accounts. One account earns interest at 9.5 percent interest and the other at 11 percent. If the total annual interest from the two investments is $2,571.75, what amount is invested in each of the accounts?

The equation consists of terms, each of which, represents an amount of interest. Interest from the 9.5 percent investment plus the interest from the 11 percent investment equals the total interest.

Interest Rate	9.5%		11%		
Amount	x	+	25,500 – x	=	$2,571.75

$$25,500 = \text{total amount invested}$$
$$x = \text{amount invested at } 9.5\%$$
$$25,500 - x = \text{amount invested at } 11\%$$

$$0.095x + 0.11(25,500 - x) = 2,571.75$$

$$0.095x + 2,805 - 0.11x = 2,571.75$$

$$-0.015x = 2,571.75 - 2,805$$

$$-0.015x = -233.25$$

$$x = \$15,550 \text{ invested at } 9.5\%$$

$$25,500 - x = \$9,950 \text{ invested at } 11\%.$$

B. Rate of Work Problems

The solution of this type of problem involves the fractional portion of the total job that is done in unit time by each participant. For example, if John can complete a job in six hours, he can do one-sixth of the job in one hour.

Example: Bob can plow a certain field in 10 hours and Melissa could plow the same field in six hours. If Bob plows for four hours, how long would it take to finish the job if Melissa and Bob both work until the plowing is finished?

In one hour Bob can do $\frac{1}{10}$ of the job; Melissa can do $\frac{1}{6}$ of the job. Bob has plowed for four hours. He has completed $\frac{4}{10}$ of the job.

Let x = hours that Bob and Melissa work together on the job

Bob's part of the job + Melissa's part of the job = the whole job

$$\frac{4}{10} + \frac{x}{10} + \frac{x}{6} = 1$$

Solving, multiply each term by the LCM of 6 and 10, which is 30:

$$\overset{3}{\cancel{30}} \cdot \frac{4}{\underset{1}{\cancel{10}}} + \overset{3}{\cancel{30}} \cdot \frac{x}{\cancel{10}} + \overset{5}{\cancel{30}} \cdot \frac{x}{\cancel{6}} = \overset{}{30} \cdot 1, \quad \text{then } 12 + 3x + 5x = 30$$

$$8x = 30 - 12$$
$$8x = 18$$
$$x = 2\tfrac{1}{4} \text{ hr.}$$

C. Problems Involving Geometric Figures

Example: The perimeter of a triangular field is 32 rods. If the second side is 3 rods longer than the first side and the third side is 2 rods longer than the second side, what is the length of each side of the field?

The perimeter of a geometric figure is the sum of the lengths of its sides.

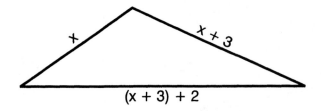

Let x = the length of the first side
x + 3 = the length of the second side
x + 3 + 2 = the length of the third side

$$x + x + 3 + x + 3 + 2 = 32$$

$$3x + 8 = 32$$

$$3x = 32 - 8$$

$$3x = 24$$

$$x = 8 \text{ rods (length of first side)}$$

---

x + 3 = 8 + 3 = 11 rods (second side)

x + 5 = 8 + 3 = 13 rods (third side)

D. General Problems

Example: Paul, Brad, and Jane are to share the net profit of $1,825 received from raising sweet corn for the early summer market. Based on the equipment furnished and the time and labor spent on the project, Brad is to get three times as much as Paul, and Jane is to get $100 less than Brad. How much should each receive?

Paul's share + Brad's share + Jane's share = $1,825

Let x = Paul's share
3x = Brad's share
3x – 100 = Jane's share

x + 3x + 3x – 100 = 1825

$$7x = 1825 + 100$$

$$7x = 1925$$

$$x = \$275 \text{ (Paul's share)}$$

$$3x = \$825 \text{ (Brad's share)}$$

$$3x - 100 = \$725 \text{ (Jane's share).}$$

Exercise 7–2

SOLVING STATED PROBLEMS

1. Three sisters: Jane, Keri, and Alice purchase a purebred riding horse. If Jane paid for one-fourth of the price of the horse, Keri paid 400 more than Jane, and Alice paid 400 more than Keri, what was the price of the horse?

_____

2. A farm and home supply center received a bill for a shipment of 246 rails and posts in the amount of $1,007.78. If rails cost $3.62 each and posts cost $5.05 each, how many rails and how many posts had been shipped?

_____

3. The Bascom Feed Store sells grain mixtures for horse feed. If cracked corn sells for 12 cents per pound, and crimped oats sells for 13 cents per pound, how many pounds of each of these grains should be mixed in order for a 25-pound bag of feed to sell for $3.08.

_____

4. A 50-foot length of ½-inch lead rope is cut into three pieces. Two of the pieces are of equal length and the third piece is 8 feet longer than each of the other two pieces. What is the length of each of the pieces?

_____

5. Jason and Marie are typing the annual report for Jepson's Feed and Seed Company. Jason could type the entire report by himself in 8 hours and Marie could type the report alone in 12 hours. How long should it take to type the report if they work together on it?

_____

6. A farmer ordered 2 tons of fertilizer containing 38 percent soluble potash. Fertilizer materials available at the blending plant are potassium chloride having 60 percent soluble potash and potassium magnesium sulfate having 22 percent soluble potash. How much of each material should be mixed to fill the order?

_____

7. Elizabeth and Eric are forepersons for bin construction crews. Elizabeth's crew can erect a large cylindrical bin in 30 hours and Eric's crew needs 45 hours of working time to do the same job. After Elizabeth's crew has worked on a bin for 10 hours, Eric's crew joins them and the two crews work together to finish the job. How long will it take the two crews to complete the construction of the bin?

_____

8. Fred spent one-third of his summer vacation at a dog obedience training school. He then tended a dog kennel for 13 days and at the end of that time one-half of his vacation was gone. How many days of vacation did he have?

_____

9. A beef farmer sold 16 steers weighing a total of 20,400 pounds and received a check for $12,879.60. If choice grade steers brought a price of $64.60 per hundredweight and good grade steers brought $57.40 per hundredweight, how many pounds of beef were sold at each grade?

_____

_____

Name_____   Date_____

10. A horse farm has 650 acres of farm land to plant in hay and oats. If the farm's feed requirements are best met if there are 180 more acres of hay than oats, how many acres of each crop should be grown?

_____

11. A feeding floor has a perimeter of 176 feet. If the width of the floor is three-eights of the length, what are the dimensions of the feeding floor?

_____

12. The Echo Exchange Club is having a banquet and members have ordered corsages for their guests. John could make all of the corsages alone in six hours. Elise could make the corsages in five hours. If John has already worked for two hours when Elise joins him, how long will it take them to finish the job working together?

_____

13. When Ellen cashed her weekly paycheck of $210.00, she received it in 10- and 20-dollar bills. The number of 20-dollar bills was three more than twice the number of 10-dollar bills. How many of each kind of bill did she get?

_____

14. Instructions for mixing concrete for a farm shop floor state: Use 3 times as much sand as cement and 1½ times as much gravel as sand. If the combined supply of sand and gravel is 195 cubic feet, how much cement is needed to mix with all of the available sand and gravel?

_____

15. Five hundred fifty pounds of milk testing 2.6 percent butterfat are to be mixed with enough rich milk to bring the test up to 3.8 percent butterfat and double the original volume. What butterfat content must the rich milk have to accomplish this?

_____

16. A ton (dry matter basis) of concentrate having a 17.6 percent digestible protein content is to be mixed for dairy cows using ground ear corn and soybean oilmeal. If ground ear corn is 4.6 percent digestible protein and soybean oilmeal is 43.8 percent digestible protein, what amount of each should be used for the concentrate?

_____

17. Labeling machine A requires six hours to label the cans of sweet corn from a day's run at the local canning factory. Labeling machine B can do the same job in three hours. After machine A has been in operation for one hour, machine B is also put into use. How much longer will the two machines have to run to complete the job?

_____

18. A rectangular dog-proof night enclosure is to be constructed for a flock of sheep. The enclosure is to be twice as long as it is wide. If the farmer has 55 meters of suitable fence and will use a barn for one of the long sides, what will the dimensions of the enclosure be?

_____

19. A security light is to be installed in a farmyard. If one-fifth of the pole is in the ground and 24 feet of the pole are above ground, what is the length of the pole?

_____

20. In a fertilizer blending plant, a mixing tank can be filled by pipe A in 42 minutes and by pipe B in 30 minutes. If pipe A is used alone for 20 minutes and then pipe B is opened, how long will it take to finish filling the tank using both pipes?

_____

Many stated or verbal problems that involve two unknown quantities often can be solved more conveniently by using two variables to form two equations that fit the conditions of the problem. We call these two equations a **system of equations** or simultaneous equations. Consider the two equations, $2x - y = 3$ and $x + y = 3$. We are looking for a value of x and a value of y that will fit both equations. By inspection, we see that $x = 2$ and $y = 1$ are the values of x and y that will make both statements of equality true.

$$2x - y = 3 \qquad\qquad x + y = 3$$
$$2(2) - 1 = 3 \qquad\qquad 2 + 1 = 3$$
$$4 - 1 = 3 \qquad\qquad 3 = 3$$
$$3 = 3$$

To solve for values of x and y algebraically, we need to eliminate one of the variables and develop an equivalent equation in one variable. We can use either the **elimination by addition method** or the **elimination by substitution method**.

A. Elimination by Addition

The general plan of solving a system of equations by the addition method is to get the coefficients of one of the variables equal but having opposite signs. The coefficients will then **add to zero** and **eliminate** that variable. Addition of the remaining terms of the system results in an equation in one variable.

Example: $2x - y = 3$
              $x + y = 3$

Add the two equations and eliminate the y-terms.

Step 1.   Add the two equations:

$$2x - y = 3$$
$$\underline{x + y = 3}$$
$$3x + 0 = 6 \quad \text{or} \quad 3x = 6$$

Step 2.   Solve the resulting equation:

$$3x = 6$$
$$x = 2$$

Step 3. To find y, substitute 2 for x in either of the two original equations:

$$x + y = 3 \qquad\qquad or \qquad\qquad 2x - y = 3$$
$$2 + y = 3 \qquad\qquad\qquad\qquad\qquad 2(2) - y = 3$$
$$y = 1 \qquad\qquad\qquad\qquad\qquad\quad 4 - y = 3$$
$$-y = -1$$
$$y = 1$$

Result: $x = 2$, $y = 1$

Example: $5x + y = 2$
$4x - 3y = 13$

Eliminate the y-terms by addition.

Step 1.   In order to have the y-terms add to zero, multiply the first equation by 3 and leave the second as is:

$$3(5x + y = 2) = 15x + 3y = 6$$

Step 2.   Add the two equations:

$$15x + 3y = 6$$
$$\underline{4x - 3y = 13}$$
$$19x + 0 = 19 \qquad or \quad 19x = 19$$

Step 3.   Solve the resulting equation:

$$19x = 19$$
$$x = 1$$

Step 4.   To find y, substitute 1 for x in either of the two original equations:

$$4x - 3y = 13 \qquad\qquad or \qquad\qquad 15x + 3y = 6$$
$$4(1) - 3y = 13 \qquad\qquad\qquad\qquad\quad 15(1) + 3y = 6$$
$$4 - 3y = 13 \qquad\qquad\qquad\qquad\qquad 15 + 3y = 6$$
$$-3y = 9 \qquad\qquad\qquad\qquad\qquad\quad 3y = -9$$
$$y = -3 \qquad\qquad\qquad\qquad\qquad\quad y = -3$$

Result: $x = 1$, $y = -3$

Example: $2x - 3y = 18$
$7x - 2y = -5$

Eliminate either the x-term or the y-term by addition. Let us eliminate the x-term.

Step 1. In order to have the x-terms add to zero, multiply the first equation by 7 and the second equation by –2:

$$7(2x - 3y = 18) = 14x - 21y = 126$$
$$-2(7x - 2y = -5) = -14x + 4y = 10$$

Step 2. Add the resulting equations:

$$14x - 21y = 126$$
$$\underline{-14x + 4y = 10}$$
$$0 - 17y = 136 \quad \text{or} \quad -17y = 136$$

Step 3. Solve the resulting equation:

$$-17y = 136$$
$$y = -8$$

Step 4. To find x, substitute –8 for y in either of the two original equations:

$$2x - 3y = 18$$
$$2x - 3(-8) = 18$$
$$2x + 24 = 18$$
$$2x = -6$$
$$x = -3$$

Result: $x = -3, y = -8$

B. Elimination by Substitution

The general plan of solving a system of equations by the substitution method is to solve one of the equations of the system for one variable in terms of the other. This expression is then **substituted** for that variable in the other equation of the system. This **eliminates** one of the variables and the resulting equation can be solved as an equation in one variable. This method works best when at least one of the variables has a coefficient of one (1).

Example: $2x - y = 3$
$x + y = 3$

Solve the second equation for x and substitute this value into the first equation.

Step 1. Solve the second equation for x:

$$x + y = 3$$
$$x = 3 - y$$

Step 2.   Substitute 3 – y for x in the first equation:

$$2(3 - y) - y = 3$$
$$6 - 2y - y = 3$$
$$-3y = -3$$
$$y = 1$$

Step 3.   To find x, substitute 1 for y in either of the original equations:

$$x + y = 3$$
$$x + 1 = 3$$
$$x = 2$$

Result: x = 2, y = 1

Example: $2x + y = 8$
$8x - 5y = 5$

Solve the first equation for y and substitute this value into the second equation.

Step 1.   Solve the first equation for y.

$$2x + y = 8$$
$$y = 8 - 2x$$

Step 2.   Substitute 8 – 2x for y in the second equation and solve for x.

$$8x - 5(8 - 2x) = 5$$
$$8x - 40 + 10x = 5$$
$$18x = 45$$
$$x = 2\frac{1}{2}$$

Step 3.   To find y, substitute 2½ for x in either of the original equations.

$$2(2\frac{1}{2}) + y = 8$$
$$5 + y = 8$$
$$y = 3$$

Result: x = 2½ , y = 3

C. Solution of Verbal Problems

Example: Two hundred fifty trays of bedding plants were shipped from a greenhouse to a garden center. The shipment consisted of petunias at $1.29 per tray and snap-dragons at $1.76. If the total bill for the plants was $360.80, how many trays of each kind of plant were shipped?

Let p equal the number of trays of petunias and s the number of trays of snap-dragons. Set up one equation involving the number of trays of plants and a second equation relating the value of the plants.

$$p + s = 250$$
$$1.29p + 1.76s = 360.80$$

Solve the system of equations by addition or substitution.

Example: The length of a rectangular field is 6 rods more than its width. If the perimeter is 52 rods, what are the dimensions of the field?

Let ℓ equal the length and w the width of the field. Set up one equation giving the relationship between the length and the width and a second equation giving the relationship of the length and width to the perimeter.

$$\ell = w + 6$$
$$2\ell + 2w = 52$$

This system of equations will be solved most easily by substitution.

D. The Pearson Square

An alternative shortcut method of solving problems involving mixtures is the **Pearson Square**. We can consider only two components, and the percent composition or the cost of each component must be known. Set up the diagram as follows:

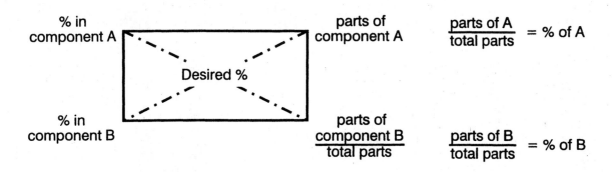

Example: (Problem 18 of Exercise 7–3) How many pounds each of 3 percent milk and 5 percent milk should be mixed to make 100 pounds of 4.5 percent milk?

Solution:

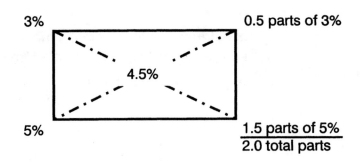

$$\frac{0.5}{2.0} = 0.25 \times 100 = 25\% \times 100\# = 25\# \text{ of } 3\%$$

$$\frac{1.5}{2.0} = 0.75 \times 100 = 75\% \times 100\# = 75\# \text{ of } 5\%$$

Example: The Newsome Feed Store sells grain mixtures for horse feed. If cracked corn sells for 13 cents per kilogram and crimped oats sells for 9 cents per kilogram, what amount of corn and oats should be mixed so that a 25-kilogram bag of the mixture would sell for $2.70?

Determine the ¢/kg of the final mixture by dividing $2.70 by 25. Final mixture is 10.8¢/kg.

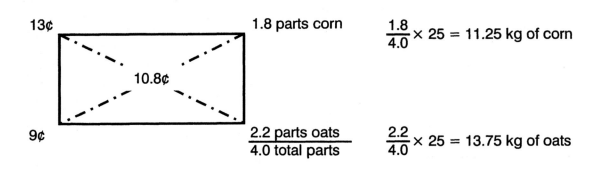

Name_____ Date_____

Exercise 7–3

SYSTEMS OF EQUATIONS

Solve the following systems of equations by the method of elimination by addition.

1. $x + 3y = 7$
   $x - 5y = -1$

2. $x + 2y = 3$
   $-2x + y = 14$

3. $2x + 5y = 10$
   $-3x + y = -15$

4. $3x + 8y = 4$
   $x - 2y = -1$

5. $2a - 5b = 8$
   $5a + 7b = 7$

Solve the following systems of equations by substitution.

6. $2x - y = 17$ and $2x - 7y = -1$

7. $4x - y = -4$ and $10x - 2y = -7$

8. $3x - 8y = 10$ and $x - 4y = 3$

9. $x + 0.5y = 1$ and $2x + 2y = 1$

10. $0.1x + 0.3y = 5$ and $3x - y = 0$

Solve the following problems by setting up a system of equations in two variables.

11. Cranstrom's Farm and Home Store mixes its own grass seed. How many pounds of Bermuda, which sells for $3.25 per pound, and of fescue, which sells for $2.75 per pound, should be mixed to make a 25-pound bag of grass seed that sells for $40.75 per bag?

_____

12. At a recent plant sale, the Hort Club sold a total of 54 African violets and gloxinias. The African violets sold for $3.00 and gloxinias sold for $7.00. The total receipts from the sale were $262.00. How many of each kind of plant were sold?

_____

13. Urea, which contains 45 percent nitrogen by weight, and ammonium sulfate, which contains 20 percent nitrogen by weight, are to be blended for an application of nitrogen. What amounts of each should be mixed in a ton of fertilizer if the desired nitrogen content is 36 percent?

_____

14. If four avocados and three grapefruit cost $5.12 and if two avocados and five grapefruit cost $4.30, then what is the price of a single avocado and a single grapefruit?

_____

15. A technician is to mix up a liter of 2.75 percent tincture of iodine for a sheep farmer to use as a disinfectant when docking lambs. Available is a supply of 7 percent tincture of iodine and 1 percent tincture of iodine. How many milliliters of each available tincture of iodine should be used? (One liter contains 1,000 milliliters.)

_____

16. A farmer wishes to invest a part of $7,500 in stocks earning 10 percent dividends and the remainder in tax-exempt bonds earning 7 percent. How much should be invested in stocks and bonds if the farmer wishes to receive an average of 8 percent return on the total invested?

_____

17. A crop farmer sold 7 truckloads of shelled corn and 5 truckloads of soybeans to the local elevator. The total amount of corn and soybeans was 6,120 bushels. If the corn sold for $2.95 per bushel and soybeans for $6.70 per bushel and the total receipts were $25,962.75, how many bushels of each crop were sold?

_____

18. How many pounds each of 3 percent milk and 5 percent milk should be mixed to make 100 pounds of 4.5 percent milk?

_____

19. How many pounds each of two kinds of feed, one worth 3 cents per pound and one worth 4.5 cents per pound, should be used to make 200 pounds of feed worth $3.75 per hundred pounds?

_____

20. How many pounds each of a grain mixture testing 10 percent digestible protein and silage testing 1.2 percent digestible protein are needed to supply a growing dairy animal with 25 pounds of feed containing 1 pound of digestible protein?

_____

Solve the following problems using the Pearson Square Method.

21. Grass seed A is worth $1.00 per pound and seed B is worth $1.35 per pound. How much of seeds A and B should be used to make a 50-pound bag of grass seed worth $57.00 per bag?

_____

22. How many gallons each of 2 percent butterfat milk and 4 percent butterfat milk should be mixed to produce 40 gallons of 2.5 percent butterfat milk?

_____

23. A technician wants to make up a liter (1,000 milliliters) of 42 percent glucose solution from existing supplies of 55 percent glucose and 30 percent glucose. How many milliliters of each should be used?

_____

8

RATIOS AND PROPORTIONS

Key Terms

direct proportion
distance (D) from fulcrum
distance formula
effort (E)
extremes
fertilizer analysis
fulcrum
indirect proportion
inverse proportion

lever
means
meshing gears
partitive proportion
proportion
pulley
ratio
resistance (R)
similar geometric figures

Career Connections

RETAIL FLORIST

Flowers and floral arrangements are used for decoration and gifts. A retail florist is responsible for preparing flowers for special occasions.

A florist cares for flowers on arrival, waits on customers, takes telephone orders, creates arrangements, handles money, and makes deliveries. The florist must have a strong sense of color and be creative.

Florists get on-the-job training. They may also go to specialized seminars at colleges or technical schools. Florists have technical or community college education or a college degree in horticulture.

A **ratio** expresses the relationship between numbers or quantities. To denote the comparison of the integers 3 and 5, for example, the ratio may be written with a colon,

3:5

with a division sign,

3 ÷ 5

or with a fraction bar,

$\frac{3}{5}$.

The ratio would be read, "three is to five" or "three to five." All of the suggested notations indicate that the ratio is actually a comparison of 3 and 5 by division or by their quotient. When comparing quantities with units in ratio form, the quantities must be expressed in the same unit of measurement. The ratio of 15 millimeters to 2 centimeters would not be 15:2 but 15:20 or 3:4.

Ratios may be used in a variety of ways as shown in the following examples:

1. Find the ratio of 14 to 6.

    If the numbers or quantities to be compared have a common factor, express in ratio form and reduce to lowest terms.

    $14:6 = 7:3$ or $\frac{7}{3}$.

    The whole number ratio is generally preferred, but the ratio could be expressed

    $2\frac{1}{3}:1$.

2. Find the ratio of

    $\frac{1}{2}$ to $\frac{2}{3}$.

    Write the ratio as

    $\frac{1}{2} \div \frac{2}{3}$.

This now becomes a problem in the division of common fractions. Invert the divisor and multiply.

$$\frac{1}{2} \times \frac{3}{2} = \frac{3}{4} \text{ or } 3:4.$$

3. Find the ratio of

   $$6\frac{1}{8} \text{ to } 3\frac{1}{2}.$$

   Change the numbers to improper fractions, express the ratio in the form

   $$\frac{49}{8} \div \frac{7}{2},$$

   invert the divisor, and multiply.

   $$\frac{49}{8} \times \frac{2}{7} = \frac{7}{4} \text{ or } 7:4.$$

4. Find the ratio of 2 feet to 15 inches.

   Convert 2 feet to 24 inches to make the units alike, express as a ratio, and reduce to lowest terms.

   $$\frac{24 \text{ inches}}{15 \text{ inches}} = \frac{8}{5} \text{ or } 8:5.$$

5. Find the ratio of $3.00 to 50 cents

   Convert $3.00 to 300 cents, express as a ratio, and reduce to lowest terms.

   $$\frac{300 \text{ cents}}{50 \text{ cents}} = \frac{6}{1} \text{ or } 6:1.$$

6. Find the gear ratio of two gears that have 75 teeth and 15 teeth, respectively.

   $$\frac{75 \text{ teeth}}{15 \text{ teeth}} = \frac{5}{1} \text{ or } 5:1.$$

7. Find the ratio of 5 square centimeters to 40 square millimeters.

   In the appendix, find the information that $1 \text{cm}^2 = 100 \text{ mm}^2$.

   $$5 \text{ cm}^2 = 500 \text{ mm}^2$$

   $$\frac{500 \text{ mm}^2}{40 \text{ mm}^2} = \frac{25}{2} \text{ or } 25:2.$$

   Note that in examples 4, 5, 6, and 7, the units of the ratio cancel each other and the ratio becomes an abstract number having no units.

Exercise 8–1

RATIOS

Express each of the following relationships as a ratio of whole numbers in lowest terms using either the fraction bar or the colon.

1. 24 to 132 = _____

2. 15 inches to 5 feet = _____

3. 162 to 18 = _____

4. A dime to a quarter = _____

5. 12 to $\frac{1}{3}$ = _____

6. 18 ounces to 1 pound = _____

7. $\frac{1}{8}$ to 36 = _____

8. 15 minutes to 1 hour = _____

9. 3 to 13 = _____

10. 30 millimeters to
    1 centimeter = _____

11. $\frac{1}{4}$ to $\frac{1}{6}$ = _____

12. 1 quart to 2 gallons = _____

13. $\frac{3}{8}$ to $1\frac{5}{16}$ = _____

14. 2 inches to 1 foot = _____

15. $12\frac{3}{4}$ to $1\frac{1}{2}$ = _____

16. 4 hours to 2 days = _____

17. 0.3 to 1.2 = _____

18. 2 quarts to 1 pint = _____

19. $\frac{5}{6}$ to $\frac{2}{3}$ = _____

20. $1.25 to 25 cents = _____

An expression of quality between two ratios is called a **proportion**. The statement of equality

$$\frac{1}{4} = \frac{3}{12}$$

is a proportion. The proportion could also be written with colons as 1:4::3:12 and would read "one is to four as three is to twelve." The first term of the proportion is 1; 4 is the second term; 3 is the third term; and 12 is the fourth term. The second and third terms of the proportion are called the **means**, and the first and fourth terms are called the **extremes**. In a true proportion, **the product of the means is equal to the product of the extremes**. If the quantities compared in the ratios of the proportion vary so that as one quantity increases the other also increases or as one quantity decreases the other also decreases, the relationship is called a **direct proportion**. If, however, an increase in one quantity causes a corresponding decrease in the other quantity or a decrease in one quantity causes a corresponding increase in the other quantity, the relationship is called an **indirect** or **inverse proportion**.

A proportion is a fractional equation and is most frequently solved by making use of the fact **the product of the means equals the product of the extremes**. This process is also referred to as "cross multiplication." In finding the value of x if 4 is to x as 5 is to 30, we write the proportion in equation form and solve.

$$\frac{4}{x} = \frac{5}{30}.$$

The product of the means (5 times x) equals the product of the extremes (4 times 30), so 5x = 120 or x = 24.

If fractional numbers or quantities occur in the terms of the proportion, it is often helpful to simplify each side of the equation before completing the solution of the proportion. In the solution of the proportion

$$\frac{1\frac{3}{5}}{3\frac{1}{5}} = \frac{\frac{3}{4}}{x},$$

the left-hand side becomes

$$\frac{\frac{8}{5}}{\frac{16}{5}}.$$

Invert the divisor and multiply:

$$\frac{8}{5} \times \frac{5}{16} = \frac{1}{2}.$$

The right-hand side of the equation becomes

$$\frac{3}{4x}$$

and the new proportion is

$$\frac{1}{2} = \frac{3}{4x}.$$

Cross multiplication gives us the equation $4x = 6$ or $x = 1.5$.

Exercise 8–2

PROPORTIONS

Solve for the variable in the following problems.

1. $\dfrac{x}{12} = \dfrac{5}{7}$ x = _____

2. $\dfrac{24}{N} = \dfrac{15}{25}$ N = _____

3. $\dfrac{\frac{3}{4}}{9} = \dfrac{7}{y}$ y = _____

4. $\dfrac{x}{6} = \dfrac{\frac{3}{4}}{\frac{9}{16}}$ x = _____

5. $\dfrac{\frac{1}{9}}{\frac{4}{27}} = \dfrac{3}{x}$ x = _____

6. $\dfrac{3.6}{5.4} = \dfrac{x}{2.5}$ x = _____

7. $\dfrac{58}{N} = \dfrac{24}{5}$ N = _____

8. $\dfrac{2\frac{1}{4}}{13\frac{1}{2}} = \dfrac{3\frac{1}{2}}{N}$ N = _____

Proportions can be used to divide a quantity into shares or parts that are in the same ratio to each other as a series of numbers are to each other. This is called a **partitive proportion**.

Example: "96" is to be divided into three parts proportional to the number series $1 : 4 : 7$.

Add 1, 4, and 7 to get 12 total parts, then set up a proportion with each number of the series.

$$\frac{1}{12} = \frac{x}{96} \qquad\qquad \frac{4}{12} = \frac{x}{96} \qquad\qquad \frac{7}{12} = \frac{x}{96}$$

$$12x = 96 \times 1 \qquad 12x = 96 \times 4 \qquad 12x = 96 \times 7$$

$$x = 8 \qquad\qquad\quad x = 32 \qquad\qquad\quad x = 56$$

$$8 + 32 + 56 = 96.$$

Example: Four students raised cucumbers for a pickle processor last summer and had a net profit of $6,396. Based on the amount of time spent working on the project, it was determined that students A, B, C, and D should share the profits in the ratio of

$$\frac{2}{5} : \frac{2}{3} : \frac{1}{6} : \frac{1}{2},$$

respectively. What amount should each student receive?

Find the sum of the fractions $\frac{2}{5} + \frac{2}{3} + \frac{1}{6} + \frac{1}{2}$.

$$\frac{12}{30} + \frac{20}{30} + \frac{5}{30} + \frac{15}{30} = \frac{52}{30}$$

Because all of the fractions now have the same denominator, we can use the numerators to form the ratios for the proportion.

$$\frac{12}{52} = \frac{x}{6,396} \qquad \frac{20}{52} = \frac{x}{6,396} \qquad \frac{5}{52} = \frac{x}{6,396} \qquad \frac{15}{52} = \frac{x}{6,396}$$

$$52x = 12 \times 6,396 \quad 52x = 20 \times 6,396 \quad 52x = 5 \times 6,396 \quad 52x = 15 \times 6,396$$

$$x = \$1,476 \qquad\quad x = \$2,460 \qquad\quad x = \$615 \qquad\qquad x = \$1,845$$

$$\$1,476 + \$2,460 + \$615 + \$1,845 = \$6,396.$$

Exercise 8–3

PARTITIVE PROPORTIONS

Solve using partitive proportions.

1. A soil mixture is to be made up of topsoil, peat, and vermiculite in the ratio of 5 : 3 : 1. If the soil sterilizer holds 1 cubic yard of material (1 yd³ = 27 ft³), how many cubic foot measures of each should be put into the sterilizer?

   _____

2. A concrete mixture of cement, sand, and gravel in the ratio of 1:2¾:4 is to be used for a footing. If the concrete is to be mixed on the job and approximately 31 cubic yards of concrete are to be made, what amount of each of the ingredients should be ordered?

   _____

3. A truck farmer harvested Pontiac, Kennebec, and Russet potatoes in the ratio of 5 : 6 : 3, respectively. If the total crop was 1,680 bushels, how many bushels of each variety were harvested?

   _____

4. Three corporate farmers A, B, and C were to divide profits of $32,550 in the ratio of ⅜ : ⅗ : ⅚, respectively. How much should each farmer receive?

   _____

5. A garden center ordered 600 boxes of vegetable plants from a greenhouse operator. If cabbage, celery, broccoli, and tomato plants were ordered in the ratio of 4 : 1 : 2 : 8, respectively, how many boxes of each were ordered?

   _____

In the solution of application problems involving a proportion, it is essential that each ratio be set up to give the correct relationship between the quantities compared. For problems involving a direct proportion, set up the given ratio on the left-hand side of the equation. Then set up the ratio on the right-hand side of the equation so the units are in the same relationship to each other as they are in the left-hand ratio.

Example: If rain is falling at the rate of 0.50 inch per hour, how many inches of rain will fall in 45 minutes?

$$\frac{0.50 \text{ inch}}{60 \text{ minutes (1 hour)}} = \frac{x \text{ inches}}{45 \text{ minutes}}.$$

This problem could also be set up with the same units in the same ratio having the corresponding units in the other ratio.

$$\frac{0.50 \text{ inch}}{x \text{ inches}} = \frac{60 \text{ minutes}}{45 \text{ minutes}}.$$

If the problem involves an inverse proportion, the ratio **must** be set up with the same units in the same ratio. After the left-hand ratio has been set up, the right-hand ratio is set up with the corresponding units inverted (inverse proportion).

Example: Four carpenters can build a barn in 20 days. Assuming the carpenters work at the same rate, how long would it take six carpenters to build the barn?

$$\frac{4 \text{ carpenters}}{6 \text{ carpenters}} = \frac{x \text{ days}}{20 \text{ days}}.$$

Proportions can be used to solve a variety of applied problems. Some of the frequently used relationships are:

A. Distance Formula

In the **distance formula**, $d = rt$, the distance traveled is directly proportional to the rate of travel and also directly proportional to the time traveled. The rate of travel is inversely proportional to the length of time traveled.

Example: An airplane travels 50 miles in 15 minutes. How many miles will the airplane travel in 1.5 hours?

$$\frac{50 \text{ miles}}{x \text{ miles}} = \frac{15 \text{ minutes}}{90 \text{ minutes (1.5 hours)}}$$

$$15x = 50 \times 90$$

$$x = 300 \text{ miles.}$$

Example: A plane travels at the rate of 264 miles per hour for 1 hour and 40 minutes. What is the speed of the plane if it makes the return trip in 1 hour and 20 minutes?

$$\frac{264 \text{ mph}}{x \text{ mph}} = \frac{1\frac{1}{3} \text{ hours}}{1\frac{2}{3} \text{ hours}}$$

Simplifying the right-hand side of the equation, it becomes

$$\frac{4}{5}.$$

The proportion now is

$$\frac{264 \text{ mph}}{x} = \frac{4}{5}.$$

Solving by cross multiplication, the equation is $4x = 5 \times 264$ mph, $x = 330$ mph.

B. Similar Geometric Figures

A proportion can be used to solve many types of problems involving **similar** (having the same shape) **geometric figures**.

1. If two plane geometric figures are similar, the lengths of the corresponding parts (sides, altitudes, bases, radii, diameters, circumferences, etc.) are directly proportional.

Example: A silo casts a shadow 90 feet long when a 20-foot utility pole casts a shadow 30 feet long. How tall is the silo?

The following diagrams indicate that the problem involves two similar triangles. The corresponding parts are directly proportional.

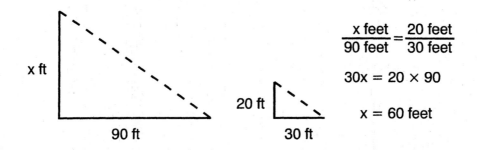

$$\frac{x \text{ feet}}{90 \text{ feet}} = \frac{20 \text{ feet}}{30 \text{ feet}}$$

$$30x = 20 \times 90$$

$$x = 60 \text{ feet}$$

2. If two plane geometric figures are similar, the areas of the figures are directly proportional to the squares of the corresponding parts.

Rectangle — $\dfrac{A_1}{A_2} = \dfrac{\ell_1^2}{\ell_2^2} = \dfrac{w_1^2}{w_2^2}$

Square — $\dfrac{A_1}{A_2} = \dfrac{s_1^2}{s_2^2}$

Triangle — $\dfrac{A_1}{A_2} = \dfrac{b_1^2}{b_2^2} = \dfrac{h_1^2}{h_2^2}$

Circle — $\dfrac{A_1}{A_2} = \dfrac{r_1^2}{r_2^2}$

Example: Square feedlot F_1 has an area two times that of square feedlot F_2 If the length of a side of feedlot F_1 is 30 meters, what is the length of a side of feedlot F_2?

$$\frac{A_1}{A_2} = \frac{s_1^2}{s_2^2}$$

$$\frac{2}{1} = \frac{(30m)^2}{s_2^2}$$

$$\frac{2}{1} = \frac{900 \text{ m}^2}{s_2^2}$$

$$2s_2^2 = 900 \text{ m}^2$$

$$s_2^2 = 450 \text{ m}^2$$

$$s_2 = \sqrt{450 \text{ m}^2}$$

$$s_2 = 21.2 \text{ m.}$$

3. If two solid geometric figures are similar, the volumes of the figures are directly proportional to the cubes of the corresponding parts.

Cube $\quad - \quad \dfrac{V_1}{V_2} = \dfrac{s_1^3}{s_2^3}$

Rectangular solid $\quad - \quad \dfrac{V_1}{V_2} = \dfrac{\ell_1^3}{\ell_2^3} = \dfrac{w_1^3}{w_2^3} = \dfrac{h_1^3}{h_2^3}$

Sphere $\quad - \quad \dfrac{V_1}{V_2} = \dfrac{r_1^3}{r_2^3}$

Example: A rectangular tank that has a length of 20 centimeters has a volume of 3,000 cubic centimeters. Find the volume of a similar tank that has a length of 60 centimeters.

$$\frac{V_1}{V_2} = \frac{\ell_1^3}{\ell_2^3}$$

$$\frac{3,000 \text{ cm}^3}{V_2} = \frac{(20 \text{ cm})^3}{(60 \text{ cm})^3}$$

$$\frac{3,000 \text{ cm}^3}{V_2} = \frac{8,000 \text{ cm}^3}{216,000 \text{ cm}^3}$$

$$V_2 \times 8,000 = 3,000 \times 216,000$$

$$V_2 = 81,000 \text{ cm}^3.$$

C. Simple Machines

Proportions may be used to solve many problems involving simple machines.

1. The **lever** is a simple machine consisting of a rigid bar supported by a pivot or **fulcrum**. If the weight of the bar can be neglected, the forces of **effort** (E) and **resistance** (R) applied to the bar are inversely proportional to their **distance** (D) **from the fulcrum**. D_E is the distance of the effort from the fulcrum and D_R is the distance of the resistance from the fulcrum. The following diagrams show three possible arrangements of the effort, resistance, and fulcrum.

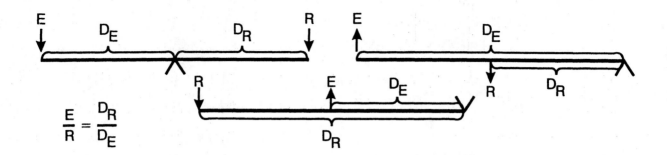

$$\frac{E}{R} = \frac{D_R}{D_E}$$

Example: A crowbar is used to lift a 220-pound block of concrete. If the fulcrum is placed 2 feet from the block, what force must be applied to the end of the crowbar, which is 5 feet from the fulcrum?

$$\frac{E}{220} = \frac{2}{5}$$

$$5E = 2 \times 220$$

$$5E = 440$$

$$E = 88 \text{ pounds.}$$

2. The speeds of rotation of machine **pulleys**, which are connected by a belt, are inversely proportional to the diameters of the pulleys.

$$\frac{R_1}{R_2} = \frac{D_2}{D_1}$$

Example: A driving pulley has a diameter of 7.5 centimeters and is rotating at 1,200 revolutions per minute. It is connected by a belt to a pulley having a 20-centimeter diameter. What is the speed of rotation of the 20-centimeter pulley?

$$\frac{R_1}{R_2} = \frac{D_2}{D_1}$$

$$\frac{7.5}{20} = \frac{x}{1,200}$$

$$20x = 7.5 \times 1,200$$

$$x = 450 \text{ rpm.}$$

3. The speeds of rotation of **meshing gears** are inversely proportional to the number of teeth on the gears.

$$\frac{R_1}{R_2} = \frac{N_2}{N_1}$$

Example: A driving gear has 35 teeth and rotates at 90 revolutions per minute. What is the speed of the driven gear that has 60 teeth?

$$\frac{R_1}{R_2} = \frac{N_2}{N_1}$$

$$\frac{90}{R_2} = \frac{60}{35}$$

$$60R_2 = 90 \times 35$$

$$R_2 = 52.5 \text{ rpm.}$$

Exercise 8–4

SOLVING STATED PROBLEMS WITH PROPORTIONS

Solve the following problems using a proportion:

1. If rain is falling at the rate of 0.50 inch per hour, how many inches of rain will fall in 45 minutes?

    _____

2. Two acres of cropland can be tilled in 45 minutes. How many acres can be tilled in 2.5 hours?

    _____

3. A diesel tractor uses 1.68 gallons of fuel per acre while pulling a plow. How many acres can be plowed with a 45-gallon tankful of fuel?

    _____

4. A centrifugal irrigation pump can deliver 3,600 gallons of water in five hours. How long would it take to pump 8,400 gallons?

    _____

5. A potato farmer has 15 acres of potatoes. If the yield from 6 acres already harvested is 2,370 bushels, how many bushels can be expected from the remaining acres?

    _____

6. If 24 apple pickers can harvest an orchard in 156 hours, in how many hours could the job be done if 12 additional pickers were hired?

   _____

7. A seed company has a corn research plot that has an area three times that of an adjoining soybean plot. The corn plot is 90 feet long and 56 feet wide. If the soybean plot has the same shape as the corn plot, what are the dimensions of the soybean plot?

   _____

8. A circular flower bed has a radius of 5 feet. If the area of the flower bed is to be quadrupled, what should be the radius of the larger flower bed?

   _____

9. A 25-foot-high shade tree can give off 32 gallons of water per day by transpiration from its leaves. What amount of water would be given off by the tree during one week?

   _____

10. If eggs cost 84 cents per dozen, what will 20 eggs cost?

   _____

11. A cargo plane travels 40 miles in 12 minutes. What is the speed of the plane in miles per hour?

   _____

12. Three carpenters can build a hay storage shed in 15 hours. Assuming that the carpenters all work at the same rate, how long would it take five carpenters to do the same job?

   _____

13. Three tablespoons of fungicide are mixed with 1 gallon of water. How much water should be mixed with 5 tablespoons of fungicide?

_____

14. If $2\frac{5}{8}$ inches on a map represents 125 miles, how many inches apart are two towns on the map that are actually 424 miles apart?

_____

15. A 150-foot roll of mesh fence weighs 46.8 pounds. What length of fence is in a partial roll that weighs 28.6 pounds?

_____

16. A homeowner has a rectangular garden that measures 14 feet by 20 feet. If the area of the garden is to be tripled, what are the dimensions of the bigger garden if it retains the same shape?

_____

17. A composter bin has a square base and height of 75 centimeters. The capacity of the bin is 187.5 liters. What is the volume of a bin having a similar shape and a height of 110 centimeters.

_____

18. A wheelbarrow is made up of a lever system having the load between the fulcrum (wheel) and the handles where the effort to lift the resistance of the load is applied. If the distance from the axle of the wheel to the end of the handles is 5 feet and 50 pounds of effort are needed to lift a load of 225 pounds, how far from the wheel is the load centered?

_____

19. A semi can make a trip to pick up a load of grain in 3 hours and 36 minutes averaging 55 miles per hour. If the loaded semi can average only 45 miles per hour, how long will the return trip take?

_____

20. There are 4.8 teeth on a large gear for each tooth on a small meshing gear. How many teeth are on the large gear if the small gear has 15 teeth?

_____

21. What size pulley should be used on a machine shaft that needs to turn at 1,750 revolutions per minute if it is belted to a 12-inch tractor pulley that is turning at 950 revolutions per minute?

_____

22. A gear on a motor shaft has 56 teeth and is turning at 1,250 revolutions per minutes. What is the speed of a driven gear that has 66 teeth?

_____

23. A motor pulley turning at 150 revolutions per minute is to be belted to a 4-inch diameter grinder pulley that needs to turn at 120 revolutions per minutes. What size pulley should be put on the motor shaft?

_____

24. A 750-pound weight is located at one end of a 7-foot bar and the pivot point is located 1.5 feet from the 750-pound weight. What force must be applied at a point 6 inches form the other end in order to lift the weight?

_____

A. Fertilizer Analysis

Fertilizers are identified by their analysis. A **_fertilizer analysis_** represents the proportion in which each of the standard ingredients nitrogen (N), phosphoric acid (P), and potash (K) are present. For example, 10-15-20 indicates the percentage of total nitrogen, phosphoric acid, and potash present in the fertilizer. The nutrients are always in the same order:

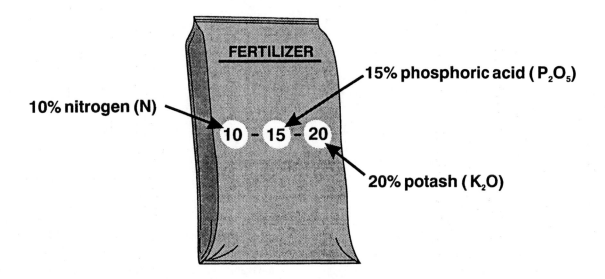

These percentages can be considered as reduced ratios.

Example: 10-15-20 has a ratio of 2-3-4. (Reduce by 5.)
            2 parts nitrogen, 3 parts phosphoric acid, 4 parts potash

Analysis	Ratio
5-10-15	1-2-3
20-20-20	1-1-1
3-6-9	1-2-3
0-10-20	0-1-2

Write the ratio for each fertilizer analysis.

Analysis	Reduce By	Ratio
1. 20-20-20	_____	_____
2. 21-7-7	_____	_____
3. 12-0-36	_____	_____
4. 15-10-30	_____	_____
5. 9-45-15	_____	_____

Fertilizers are compared on the basis of their ratios.

Example:

Analysis	Ratio
5-10-10	1-2-2
10-10-20	1-2-2

Therefore, the 5-10-10 analysis contains the three major nutrients in the same proportion as the 10-20-20, but twice (two times) as much of the 5-10-10 must be applied to obtain the same amount of nutrients as the 10-20-20 fertilizer. The ratios are the same, but the amount of nutrients in each 50-pound bag differs.

Example: How many pounds of actual nitrogen are contained in a 100-pound bag of 20-10-5 fertilizer?

Solution: 100 pounds $\times$ N = pounds of N
100 $\times$ 0.20 = 20 pounds of N

Example: How many pounds of 20-10-5 fertilizer should be purchased to apply 4 pounds of actual nitrogen to 1,000 square feet of lawn?

Solution: Divide the percentage of N into the pounds of N desired. The result is the number of pounds of fertilizer required.

4 pounds of N desired ÷ 20 = pounds of fertilizer required.
4 ÷ 0.20 = 20 pounds of 20-10-5 fertilizer required.

Example: How many pounds of each nutrient are contained in a 100-pound bag of 20-10-5 fertilizer?

Solution: 20-10-5 represents this sample analysis, which means there is 20 percent nitrogen, 10 percent phosphorus and 5 percent potassium.

Therefore:

20% of 100 = pounds of N
.20 × 100 = pounds of N
20 = pounds of N

10% of 100 = pounds of P
.1 × 100 = pounds of P
10 = pounds of P

5% of 100 = pounds of K
.05 × 100 = pounds of K
5 = pounds of K

20 lbs. of N + 10 lbs. of P + 5 lbs. of K = 35 total lbs. of nutrients

Exercise 8–5

APPLICATION EXTENSION

Fertilizer Analysis

Solve the following:

1. How many pounds of each nutrient are contained in a 100-pound bag of 5-10-15?

   _____ pounds of nitrogen

   _____ pounds of phosphorus

   _____ pounds of potassium

2. How many pounds of each nutrient are contained in a 75-pound bag of 5-10-15?

   _____

3. How many total pounds of nutrients are contained in a 100-pound bag of 5-10-15?

   _____

   In a 50-pound bag of 5-10-15?

   _____

4. How many total pounds of nutrients are contained in a 25-pound bag of 15-0-14?

   _____

5. Fertilizer recommendations required 4 pounds of nitrogen per 1,000 square feet. How many pounds of 5-10-15 are needed for each 1,000 feet?

   _____

9

SPECIAL FORMULAS

Key Terms

actual rate of interest
board feet
compound interest
drawbar horsepower
force
future amount
horsepower

maturity value
present amount
principal
power
rate of interest
work

Career Connections

FOOD INSPECTOR

Americans are concerned about the safety of their food. Many agricultural industries from meat processing plants to vegetable packaging facilities employ food inspectors to monitor the quality and safety of the products.

The responsibilities of food inspectors vary from place to place. Usually, inspectors check the sanitation of a food facility and determine the wholesomeness of the food product. Inspectors may provide written reports of their work or grade the product before distribution.

Inspectors must have high expectations. They need strong moral standards and an ability to understand the special requirements of the industry. Inspectors need specialized training and certification.

PART I. SPECIAL FORMULAS

In the solution of problems involving formulas, it is often necessary to solve for one of the factors rather than for the product. Calculations are usually easier if the formula is solved for the desired factor before the known values are substituted in the formula.

Example: Solve for S in the Drawbar Horsepower Formula:

$$HP = \frac{F \text{ (force in pounds)} \times S \text{ (speed in miles per hour)}}{375 \text{ (conversion factor)}}$$

Step 1.   Multiply each side of the equation by the common denominator, 375, and cancel.

$$375 \times HP = \frac{\cancel{375}(F \times S)}{\cancel{375}}$$

$$375\ HP = F \times S$$

Step 2.   Divide each side of the equation by the factor of S, which is F, and cancel.

$$\frac{375\ HP}{F} = \frac{\cancel{F} \times S}{\cancel{F}},\ S = \frac{375\ HP}{F}$$

Example: Solve for w in the Electricity Formula:

$$C = \frac{wtr}{1,000}$$

Step 1.   Multiply each side of the equation by the common denominator, 1,000, and cancel.

$$1,000 \cdot C = \cancel{1,000} \cdot \frac{wtr}{\cancel{1,000}}$$

$$1,000C = wtr$$

Step 2.   Divide each side of the equation by the factors of w, which are tr, and cancel.

$$\frac{1,000C}{tr} = \frac{w\cancel{tr}}{\cancel{tr}}$$

$$w = \frac{1,000C}{tr}$$

In the use of the following formulas for solving problems, solve for the desired factor in the formula, if necessary; then substitute the known values and solve the problem.

A. Maturity Value

The maturity value of a loan is the single-payment amount that the borrower must pay when the note is due. It is found by adding the principal and the interest earned. The formula is:

M = p + prt, where

  M = the maturity value

  p = the principal (amount of the loan)

  r = the rate of interest

  t = time in years

B. Horsepower

Definitions:

1. **Force** — that which produces or prevents motion, a push or a pull; frequently involves weight.

2. **Work** — the product of force (weight) and a distance.

   W (work) = F (force) × s (distance), W = Fs

   The unit of work is the foot-pound (ft-lb). A force of 1 pound acting through a distance of 1 foot does 1 **foot-pound** of work.

3. **Power** — the time rate of doing work.

   $$P \text{ (power)} = \frac{W \text{ (work)}}{t \text{ (time)}}, \quad P = \frac{FS}{t}$$

4. **Horsepower** — the basic unit of power.

   One horsepower is doing work at the rate of 550 foot-pounds of work per second or 33,000 foot-pounds of work per minute. If time is given in seconds, use the formula:

$$HP = \frac{Fs}{550 \text{ ft-lb} \times \text{sec}}$$

If the time is given in minutes, use the formula:

$$HP = \frac{FS}{33,000 \text{ ft-lb} \times \text{min}}$$

The term **drawbar horsepower** is used when pounds of draft of field machinery and the speed of field travel are considered. The formula used for drawbar horsepower is:

$$HP \text{ (drawbar)} = \frac{F \text{ (draft in pounds)} \times S \text{ (speed in mph)}}{375 \text{ (a conversion unit)}},$$

$$\text{or } HP = \frac{F \times S}{375}$$

C. Board Feet

Dimension lumber is measured in units called **board feet**. One board foot is a piece of lumber 1 inch thick, 12 inches wide, and 12 inches long. The formula generally used is:

$$\text{Board feet} = \frac{\text{number of pieces} \times T \times W \times L}{12}, \text{ in which}$$

T = thickness in inches,

W = width in inches, and

L = length in feet

D. Cost of Electricity

The kilowatt-hour is the electrical energy supplied to an electrical circuit or appliance in one hour by 1,000 watts of power. Electric light bulbs and appliances are rated in terms of watts consumed per hour of operation. Motors are generally rated in horsepower. **One horsepower equals 746 watts.**

To find the cost of operating electrical lights or appliances for a given period of time, the following formula may be used.

$$C = \frac{wtr}{1,000}, \text{ in which}$$

C = cost in cents for operating an appliance,

w = wattage of the appliance,

t = time in hours the appliance operates, and

r = cost in cents per kilowatt-hour

E. Actual Rate of Interest on Installment Plans

When merchandise is purchased on the installment plan, the customer usually makes a down payment and then pays back the remainder of the cost on a weekly or monthly basis. The total price of installment buying usually carries with it a service or carrying charge that the customer pays for the privilege of paying the bill in installments. The service charge is actually an interest charge. Many customers do not realize that the high rate of interest they are paying on an installment plan is calculated on the original balance even though part of the principal is being paid back with each payment. The formula for calculating the actual rate of interest on installment plans is:

$$R = \frac{2MI}{P(N + 1)} \times 100, \text{ where}$$

R = true rate of interest in per cent

M = number of pay periods in a year (monthly, M = 12; weekly, M = 52)

P = principal of loan or unpaid balance (cash price less any down payment)

I = interest in dollars (total amount to be paid in excess of unpaid balance)

N = actual number of payments made, not including the down payment

Example: A portable generator sells for a cash price of $459. If purchased on an install-ment plan, the terms call for a $50 down payment and 12 monthly payments of $38 each. What actual rate of interest is paid on the installment plan?

M = 12 (payments are made monthly)
P = $459 – $50 = $409
I = 12 × $38 = $456 – $409 = $47
N = 12

$$R = \frac{(2)(12)(47)}{409(12 + 1)} = \frac{1,128}{5,317} = 0.2121 \times 100 = 21.21\%$$

F. Compound Interest

In planning for a future purchase or for retirement, it is helpful to compute the value money will have if it is invested at a fixed interest rate and compounded at certain periods over a given length of time. Compounding accumulates larger amounts of money than simple interest rates. This calculation can be easily performed with a business or scientific calculator (or any calculator that has a "y^x" button). The formula is:

$$A = P(1 + r \div n)^{Tn}$$

    A = future amount

    P = present amount

    r = interest rate

    n = number of times per year the amount will be compounded

    T = time in years that money will be invested

Example: $5,000 is invested for 6 years at 5 percent compounded quarterly

    P = $5,000
    r = 5% = 0.05
    n = 4 (quarterly is four times per year)
    T = 6 years

so

$$A = 5,000(1 + 0.05 \div 4)^{6 \cdot 4}$$

Step 1.   Following the order of operations from Chapter 1, divide

$$0.05 \div 4 = 0.0125$$

$$A = 5,000(1 + 0.0125)^{6 \cdot 4}$$

Step 2.   Add one (1)

$$1 + 0.0125 = 1.0125$$

$$A = 5,000(1.0125)^{6 \cdot 4}$$

Step 3.   Calculate the exponent

$$6 \times 4 = 24$$

Step 4.   Use the "y^x" button on your calculator to find the power.

$$1.0125^{24} = 1.347$$

$$A = 5,000(1.347)$$

Step 5.   Multiply by 5,000

$$5,000 \times 1.347 = 6,735 \quad \text{so, } \$6,735 \text{ is the future amount.}$$

Example: \$9,200 invested at 6 percent compounded monthly for three years.

$$A = 9,200(1 + 0.06 \div 12)^{3 \cdot 12}$$

$$= 9,200(1 + 0.005)^{3 \cdot 12}$$

$$= 9,200(1.005)^{3 \cdot 12}$$

$$= 9,200(1.005)^{36}$$

$$= 9,200(1.197)$$

$$= 11,012.40 \quad \text{so, } \$11,012.40 \text{ is the future amount.}$$

Exercise 9–1

SPECIAL FORMULAS

Use the Maturity Value Formula to solve problems 1 through 5.

1. Find the maturity value of a $19,400 note at 8.5 percent ordinary interest, dated March 16 and due on June 30.

   _____

2. Find the principal that would yield a maturity value of $1,575.75 after 18 months at 8 percent ordinary interest. The formula solved for p is:

   $$p = \frac{M}{1 + rt}$$

   _____

3. The milk-cooling facilities of a dairy-goat farm were in need of remodeling. If the owner borrowed money at 9 percent ordinary interest and paid off the loan with a check for $8,923.22 (including the interest) after a period of one year and nine months, how much money was borrowed for the renovation?

   _____

4. A truck gardener borrowed $824.40 to cover planting-time expenses. What rate of ordinary interest had been charged if the maturity value of the 135-day loan was $855.73?

   _____

5. On June 15, a sodding company borrowed $3,786.65 to pay for a new tractor. What total amount did the sodding company have to pay on the due date of the loan, October 5, if the rate of ordinary interest was 9.5 percent?

   _____

Use the Horsepower Formulas to work problems 6 through 11.

6. A fertilizer blending plant has a conveyor having the capacity to lift 2 tons of fertilizer materials per minute to a height of 41¼ feet. What is the horsepower rating of the motor capable of running the conveyor?

_____

7. A centrifugal water pump is delivering 12 gallons of water per minute from a well 220 feet deep. What horsepower is expended by the motor driving the pump? One gallon of water weighs 8.34 pounds.

_____

8. An auger conveyor is powered by a 2-horsepower engine. How long should it take to unload a 110-bushel load of soybeans (60 pounds per bushel) into a 25-foot-high bin?

_____

9. A ½-horsepower motor runs a bale hoist. To what height can eight 60-pound bales be lifted in one minute?

_____

10. Find the drawbar horsepower of a tractor traveling 4.8 miles per hour pulling a 24-foot field cultivator requiring 275 pounds of draft per foot.

_____

11. A tractor is rated to have 68 horsepower at the drawbar. If the tractor is used to pull a plow requiring 5,220 pounds of draft, at what speed can the tractor travel?

_____

Name_____ Date _____

Use the Board Foot Formula to solve problems 12 through 15.

12. A do-it-yourself sale catalog advertises 2" × 6" × 16' pieces of lumber for $9.42 each. What is the price per board foot?

_____

13. Plans for a workbench in a farm shop call for the following pieces of lumber:

   1 piece of 2" × 8" × 12'
   1 piece of 2" × 6" × 14'
   6 pieces of 1" × 10" × 8'
   4 pieces of 4" × 4" × 8'

   At the price of $635 per 1,000 board feet, what will the lumber for the workbench cost?

_____

14. Rustic 5" × 7" × 8' treated landscape timbers are selling for $6.66 each. Find the price that would be charged per 1,000 board feet of the treated timber.

_____

15. A barn remodeling job requires the fir lumber listed below.

   12 pieces of 2" × 4" × 14'
   2 pieces of 2" × 6" × 14'
   6 pieces of 1" × 10" × 14'
   4 pieces of 1" × 8" × 10'
   2 pieces of 1" × 4" × 20'

   If the fir lumber is selling for $1,776 per 1,000 board feet, find the cost of the lumber.

_____

_____
255

Use the Cost of Electricity Formula to solve problems 16 through 19.

16. Each year an FFA chapter raises pheasants for release around their county. An incubator holding 130 pheasant eggs is run for three weeks, with the 150-watt heating element operating an average of 12 hours per day. At 7.1 cents per kilowatt-hour, what is the cost of electricity for the incubation period?

_____

17. A low-pressure sodium lamp is used for a security light on a farmstead. The light is on for an average of 14 hours each day during the month of December. If the light bill for the security light for December is $1.80 at a 6.9 cent per kilowatt-hour rate, what is the wattage of the lamp?

_____

18. One of the ventilating fans in a dairy barn runs continually during the winter season. If the fan has a ¼-horsepower motor, what does it cost to operate the fan for the month of January? Rate is 7 cents per kilowatt-hour.

_____

19. A 1½-horsepower motor is used at a bin site for an average of 10 hours per day for a two-week period. At 7.2 cents per kilowatt-hour, what does it cost to operate the motor for the two-week period?

_____

Use the Actual Rate of Interest Formula to solve problems 20 through 23.

20. A farm supply store advertises a PTO-driven sprayer pump for $170 cash for $15 down and $5.50 per week for 30 weeks. Find the actual rate of interest on the installment plan.

_____

21. A used ¾-ton pickup truck is priced at $4,200. The car sales agency offers installment terms of $600 down and $120 per month for 36 months. What is the actual rate of interest on the plan?

_____

22. A rabbit hutch for a large doe and her litter sells for a cash price of $37.50 or $8.00 down and six monthly payments of $5.25. What is the actual rate of interest on the installment plan?

_____

23. A motor-driven honey extractor has a cash price of $375. The manufacturer offers credit terms, which include a carrying charge consisting of 8 percent of cash price (not a down payment on the cash price), and payment in full in six equal monthly payments. What is the actual rate of interest for the credit plan?

_____

Use the Compound Interest Formula to solve problems 24 through 27.

24. $10,500 is invested at 7.5 percent compounded semi-annually for five years.

_____

25. $20,000 is invested at 8 percent compounded daily for 30 years.

_____

26. $31,750 is invested at 6.2 percent compounded quarterly for six years.

_____

27. $6,300 is invested at 7.8 percent compounded monthly for 18 months.

_____

10

MEASUREMENT

Key Terms

degree
denominate numbers
English system

Growing Degree Days (GDD)
metric system
unit equations

Career Connections

ADVERTISING SPECIALIST

A product has no value if no one will buy it. An advertising specialist develops ways to convince consumers to buy products.

The specialist may write advertisements for newspapers and magazines or create radio or television ads. Some advertising specialists set up displays at fairs or shows.

Advertising specialists need a college degree in agribusiness, marketing, or communications. Learning continues on-the-job by working with a more experienced person. Advancement is based on assuming responsibility, being productive, and doing a good job. Communication skills and the ability to relate to people are very important.

A great many of the everyday uses of mathematics involve units of measurement. A great variety of tools and devices, such as rulers, protractors, balances, etc., are used to make measurements. When units are attached to numbers, the numbers are called **denominate numbers**.

There are two basic systems of measurement in the United States, the customary **English system** and the **metric system**. The metric system has the advantage of being a completely decimal system, but there is little momentum to make the changeover. However, as our country imports and exports more products and as we move toward a global economy, agricultural specialists need greater familiarity with the metric system. The adoption of metrics may take a long time, but workers in agriculturally oriented occupations and industries need to be able to function mathematically with both systems.

A. Equivalent Measurements

The tables of measurements found in Section 2 of the Appendix give equivalents that are used to make conversions within each system and between systems. The equivalents are sometimes called **unit equations** because they are equal to each other and are equal to one (unit). A convenient method for using unit equations to make conversions is shown in the following examples. This method gives assurance that the given numbers are correctly used as multipliers and divisors. Simply arrange the unit equation as a fraction so that the given units will cancel and the desired units remain.

Example: Change 28 miles per gallon to kilometers per liter.

$$\frac{28 \text{ miles}}{1 \text{ gallon}} \times \frac{1 \text{ gallon}}{3.785 \text{ liters}} \times \frac{1.6093 \text{ kilometers}}{1 \text{ mile}} = 11.90 \text{ km/L}$$

Example: Change 20 ounces to milligrams.

$$20 \text{ ounces} \times \frac{28.35 \text{ grams}}{1 \text{ ounce}} \times \frac{1,000 \text{ milligrams}}{1 \text{ gram}} = 567,000 \text{ mg}$$

B. Basic Operations with Denominate Numbers

1. Addition

Step 1.   Write the denominate numbers in column form with like units in the same column and largest units at the left.

Step 2.   Add the denominate numbers.

Step 3.   Express the sum in the largest possible units.

Example: Add 3 yd 2 ft 9 in. + 1 yd 1 ft 11 in. + 2 yd 7 in.

3 yd	2 ft	9 in.		6 yd	=	6 yd		
1 yd	1 ft	11 in.	or	3 ft	=	1 yd		
2 yd		7 in.		27 in.	=		2 ft	3 in.
6 yd	3 ft	27 in.				7 yd	2 ft	3 in.

2. Subtraction

Step 1.   Write the denominate numbers in column form with like units in the same column with the largest units at the left.

Step 2.   Starting with the smallest unit, examine the top numbers to make sure the top number is larger than the bottom number. If the bottom number is larger, "borrow" one unit from the number immediately to the left. Combine like units.

Step 3.   Subtract the denominate numbers.

Step 4.   Express the difference in the largest possible units.

Example: Subtract 6 gal 2 qt 1 pt – 4 gal 3 qt 2 pt

6 gal	2 qt	1 pt		5 gal	5 qt	3 pt
– 4 gal	3 qt	2 pt	or	– 4 gal	3 qt	2 pt
				1 gal	2 qt	1 pt

Name_____ Date_____

ENGLISH AND METRIC LINEAR MEASUREMENTS AND FORMULAS

Refer to the Formulas for Measurement (Appendix, Section 1) and Tables 1, 2, and 3 (Appendix, Section 2), English and Metric Linear Measurements and Conversions.

Add and simplify:

1. 6 yd 1 ft 11 in. + 3 yd 2 ft 3 in. + 1 yd 9 in.   _____

2. 4 yd 2 ft 10 in. + 2 yd 1 ft 6 in. + 6 yd 2 ft 2 in.   _____

Subtract:

3. 3 yd 1 ft 8 in. – 1 yd 2 ft 10 in.   _____

4. 2 yd 1 ft 11 in. – 1 yd 2 ft 7 in.   _____

Using a ruler having both English and metric units, measure the lines in problems 5 and 6.

5. _____   (to nearest cm and ¼ in.)

_____ cm, _____ in.

6. _____   (to nearest 0.1 cm and ⅛ in.)

_____ cm, _____ in.

Without converting units or taking measurements, make estimations and select the correct unit: mm, cm, or m.

7. A dairy cow has a height of 120 _____ at the withers.

8. An ash tree may reach a height of 17 _____.

9. The diameter of a tractor's steering wheel is 40 _____.

10. The height of a soybean plant is 90 _____.

11. Baby pigs are given iron shots using a syringe having a needle with a diameter of 1 _____.

Make the following conversions:

12. 2 ft 3 in. = _____ yd

13. 4,155 yd = _____ mi

14. _____ mi = 685 ft

15. 0.5 dm = _____ mm

16. _____ yd = 155 m

17. 150 cm = _____ in.

18. 25 km = _____ mi

19. 6'8" = _____ m

20. _____ km = 758 dkm

21. Using a ruler, measure the radius of the circle in Figure 10–1 to the nearest $\frac{1}{8}$ inch and nearest 0.1 centimeter. Find the circumference of the circle, using both the English and metric units.

_____ in., _____ cm

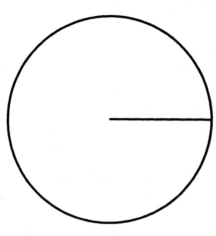

Figure 10–1

22. Using a ruler, measure each of the sides of the triangle in Figure 10–2 to the nearest $\frac{1}{8}$ inch and nearest 0.1 centimeter. Find the perimeter of the triangle in centimeters and inches.

_____ in., _____ cm

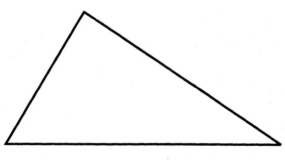

Figure 10–2

23. The perimeter of a square is 116 meters. What is the length of a side?

_____

_____

24. What is the diameter of a circular flower bed having a circumference of 15.3 meters?

_____

25. What decimal fraction of a mile do the horses run in a 6½-furlong race?

_____

26. The desirable tractor speed for a spraying operation is 5 miles per hour. To check the speed, a distance of 660 feet is measured and marked. How many minutes should it take the tractor to travel 660 feet if its speed is correct?

_____

27. A hexagon-shaped nut is shown in Figure 10–3. To the nearest whole millimeter, what is the size of the nut?

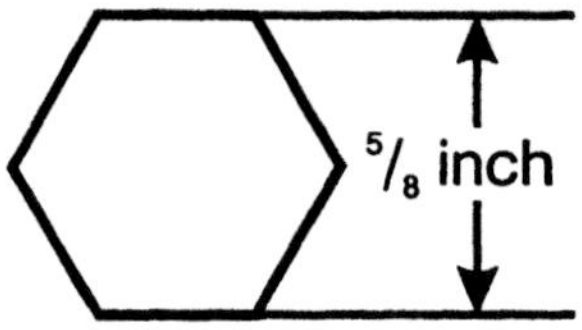

⁵/₈ inch

Figure 10–3

_____

28. Where laying hens are kept in litter-floor houses, it is recommended that 8 inches of roost be allowed for each bird. How many feet of roost are needed for a 2,100-hen laying flock?

_____

29. If 17,424 feet of a 30-inch crop row equals 1 acre, how many plants per foot of row would give a plant population of 148,104 plants per acre?

_____

30. The heartgirth measurement of a 1,250-pound steer is 79.5 inches. What is the steer's heartgirth in centimeters?

_____

31. A bale of twine (two balls) contains 9,000 feet of twine and cost $20.25. If two strands each 120 inches long are required to tie a hay bale, (a) how many hay bales can be tied with a bale of twine, and (b) what is the cost of twine per bale of hay?

a. _____

b. _____

32. A rectangular garden has a perimeter of 118 feet. If the garden is 24 feet wide, what is its length?

_____

33. A riding arena has the dimensions shown in Figure 10–4. What is the distance around the outer edge of the arena?

_____

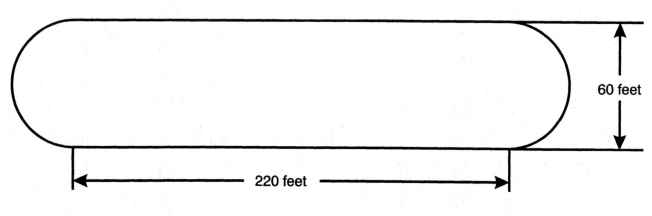

220 feet

60 feet

Figure 10–4

34. The odometer of truck A gives the distance traveled in miles while truck B's odometer gives the distance in kilometers. Truck A made a trip of 1,675 miles and truck B made a trip of 2,348 kilometers. (a) Which truck made the longer trip, and (b) how many kilometers longer?

a. _____

b. _____

35. Find the metric length (meters) of a sunflower field having rows 120 rods long.

_____

36. A barberry hedge is to be planted around the edge of the landscaped area shown in Figure 10–5. If the bushes are to be planted 45 centimeters apart, how many bushes will be needed?

_____

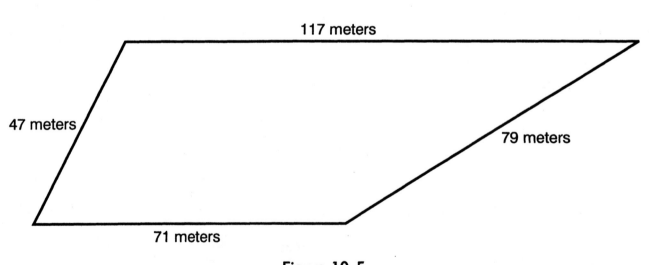

Figure 10–5

37. The plan for the concrete floor of a cylindrical bin is shown in Figure 10–6. Bolts are to be set in the concrete to secure the bin to the floor. If bolts are to be placed on the circumference at 26-inch intervals, how many bolts are needed?

_____

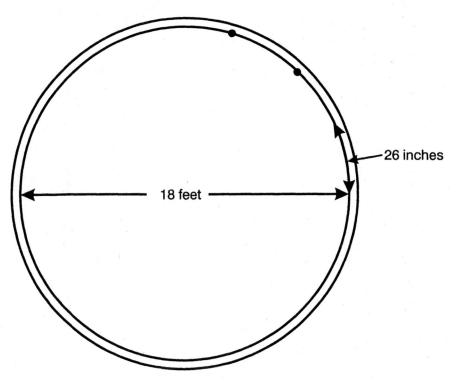

26 inches

18 feet

Figure 10–6

38. Rectangular pasture A is 225 meters long and 178 meters wide. Pasture B is 660 feet long and 435 feet wide. If each pasture is to be enclosed by a fence, (a) which pasture requires more fence, and (b) what additional length (in feet) is needed?

a. _____

b. _____

39. Metal braces are to be installed on the sliding door of a machine storage building as shown in Figure 10–7. What length of brace is needed?

_____

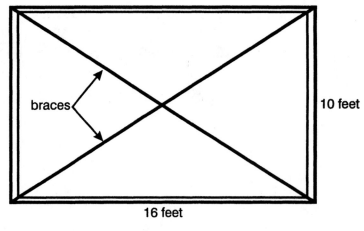

16 feet

10 feet

braces

Figure 10–7

40. Find the perimeter of the irregular plot of land shown in Figure 10–8.

_____

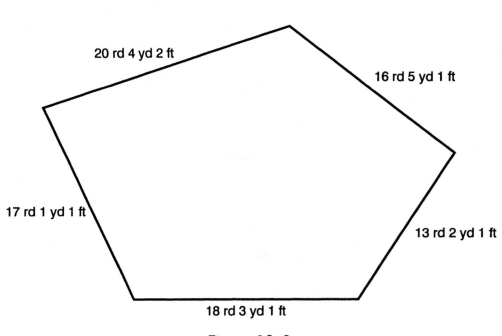

20 rd 4 yd 2 ft

16 rd 5 yd 1 ft

17 rd 1 yd 1 ft

13 rd 2 yd 1 ft

18 rd 3 yd 1 ft

Figure 10–8

41. A pasture is 42 rods long and 25 rods wide. If the pasture is to be enclosed by a five-strand barbed-wire fence, how many 80-rod rolls of barbed wire will be needed? Add an additional roll for a partial roll.

_____

42. A half-mile race track is to be constructed on a rectangular plot of land as shown in Figure
     10–9. If the ends are semi-circles, what length of the plot will need to be used for the track?
     (Use the center of the track for distance. Add one-half the width of the track to each end for
     total length of plot needed.)

_____

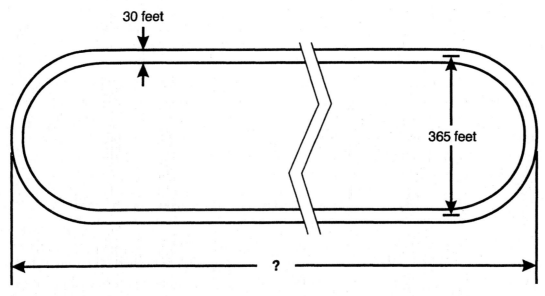

30 feet

365 feet

?

Figure 10–9

Exercise 10–2

ENGLISH AND METRIC AREA MEASUREMENTS AND FORMULAS

Refer to the Formulas of Mensuration (Appendix, Section 1) and Tables 4, 5, and 6 (Appendix, Section 2), English and Metric Area Measurements and Conversions.

Add:

1. $1{,}720$ in.^{2} + 152 in.^{2} = _____ ft^2

2. 275 ft^2 + 130 ft^2 = _____ yd^2

Subtract:

3. 4 A – 430 rd^2 = _____ rd^2

4. 625 in.^{2} – 337 in.^{2} = _____ ft^2

Without measuring or making conversions, estimate the correct response to the following statements using a, m^2, cm^2, or mm^2.

5. A backyard vegetable garden would have an area of 0.5 _____.

6. A heavy saddle blanket would have an area of 1.2 _____.

7. The cross-sectional area of the opening at the end of a garden hose would be about 450 _____.

8. The area of the bottom of a cage for a laying hen is about 0.2 _____.

9. The area of the rib eye of good steak is about 80 _____.

Make the following conversions:

10. 120.0 A = _____ m^2 11. 6.0 m^2 = _____ cm^2

12. 196 in.² = _____ ft²

13. 195 yd² = _____ ft²

14. 850.0 m² = _____ a

15. 120 rd² = _____ ft²

16. _____ cm² = 67.0 in.²

17. 76.87 m² = _____ rd²

18. 87.0 m² = _____ ft²

19. 675 m² = _____ yd²

20. 75.0 A = _____ ha

21. A flower bed having an area of 19.625 square meters is to be planted with red salvia plants. If each plant requires 625 square centimeters of space, how many plants are needed?

_____

22. It is recommended that a lawn fertilizer be applied at the rate of 50 pounds per 1,000 square feet. How much fertilizer is needed for a lawn having an area of 280 square meters?

_____

23. If 21 kernels of corn found on the ground on an area 30 inches by 30 inches represent a harvest loss of 1 bushel per acre, what is the loss if 21 kernels are found on an area 60 inches by 60 inches?

_____

24. An experimental material for floors in horse stalls costs $19.50 per square yard. If the area of a tie stall is 58.5 square feet, what would the material for this stall cost?

_____

_____

25. The recommended rate of fertilizer application is 400 pounds of 6-12-24 fertilizer per acre. At this rate, what amount of fertilizer should be spread on a garden having an area of 5,445 square feet?

_____

26. A farmer sold the acreage of highway frontage shown in Figure 10–10 to a realtor for $6,200 per acre. If the realtor divided the acreage and sold lots as shown for $9,520 each, what amount of profit did the realtor make on the transaction?

_____

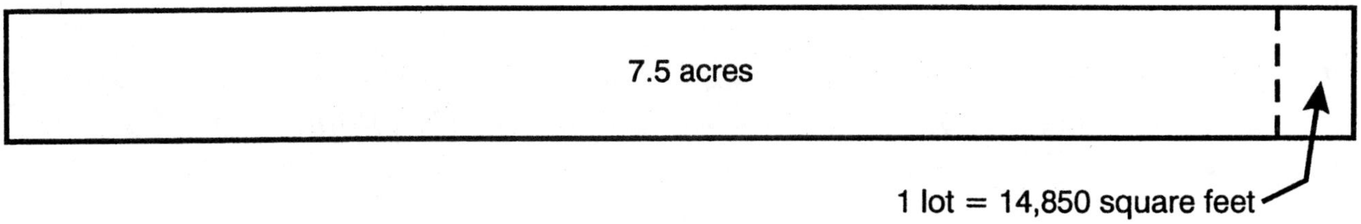

Figure 10–10

27. A barn paint, advertised to cover 375 square feet of wall surface per gallon, costs $14.89 per gallon. If a barn has 2,625 square feet of wall to be painted, what will the paint for the barn cost?

_____

28. A mature lamb's-quarters plant can produce 72,450 seeds. A badly infested soybean field has two mature lamb's-quarters plants per square yard. If viability (germination) is 55 percent, 98 percent of the seedlings will be killed by herbicide and 95 percent of the remaining plants will be killed by cultivation, what is the potential lamb's-quarters population per square yard in the next year's crop?

_____

29. A crop farmer hires custom combining for soybeans.

   a. Which of the following proposals is the better deal:

      (1) $1,384.75 for combining 95.5 acres, or
      (2) $1,472.25 for combining 39.5 hectares?

   _____

   b. By what amount in dollars per acre is one price better than the other?

   _____

30. Floor tile are to be installed in the surgical lab of a veterinary clinic. What would be the cost of the tile for the 12-foot 6-inch by 8-foot 4-inch floor if each 5-inch by 5-inch tile costs 85 cents?

   _____

31. A triangular field has two sides that meet at right angles as shown in Figure 10–11. Using the measurements shown, find the area of the field.

   _____

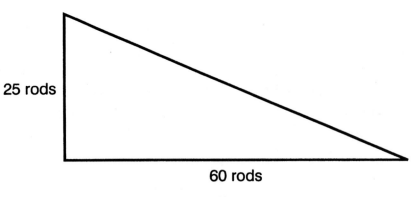

25 rods

60 rods

Figure 10–11

32. A corn combine, which harvests six 30-inch rows, is used to make a corn yield test. If 1.25 acres are required for the test, how far must the combine go into the field?

_____

33. A new road construction project takes a strip of cropland 8 rods wide and 160 rods long from the side of a farm. How many acres does the roadway remove from production?

_____

34. If the recommended area for a tie stall for horses is 52 square feet and the width is to be 5 feet 3 inches, what should be the length of the stall?

_____

35. A variety of apple trees requires a growing area 14 feet by 22 feet. How many trees can be planted in a 14-acre orchard?

_____

36. Two hundred ten feet of steam pipe, 2 inches in diameter, are used to heat a greenhouse. Find the square feet of radiation surface on the pipe.

_____

37. A landscaped lawn has the shape and dimensions shown in Figure 10–12. How many rolls of sod (18 inches by 6 feet) are needed to sod the lawn?

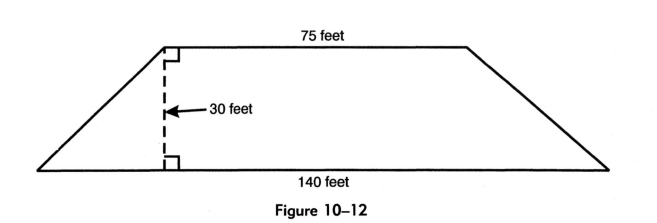

75 feet

30 feet

140 feet

Figure 10–12

_____

38. Find the lateral area of a steel grain bin having a 24-foot diameter and a wall height of 20 feet.

_____

39. Rye grain has a bushel weight of 56 pounds and each pound of rye has approximately 18,000 seeds. A fall seeding of rye was made at the rate of 1¼ bushels per acre. If the seed has a 96 percent germination, there was a 3 percent mortality of the seedlings before the winter snows came, and 15 percent of the seedlings succumbed to winterkill, how many plants per square yard could be expected in the rye field in the spring?

_____

40. The owner of the barn in Figure 10–13 wants to estimate the amount of paint needed for the building. If 1 gallon of paint is expected to cover 425 square feet of surface, how many full gallons of paint will be needed?

_____

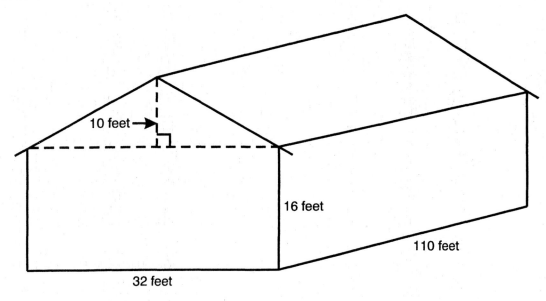

Figure 10-13

41. A tractor is pulling a 33-foot field cultivator at 4.6 miles per hour.

    a. How many acres could be tilled per hour?

    _____

    b. If turning at the ends of the field reduces the field efficiency by 12 percent, how many acres could actually be tilled per hour?

    _____

42. Ten samples of alfalfa-brome hay are taken with a forage ring that measures 1/5,000 of an acre. The total sampling from 10 different parts of the field weighs 20 pounds at 80 percent moisture. What is the potential yield of hay in tons from an acre of this field if the forage is baled at 27 percent moisture? (Note: The dry matter in the forage remains the same regardless of the moisture content.)

    _____

43. The boom of a crop sprayer is 80 feet wide. If the tractor is traveling 5 miles per hour, how many acres can be sprayed per hour?

    _____

44. The aeration system for effluent from a canning factory has a rotating sprinkler reaching a distance of 1,307 feet from the pivot point.

   a. What acreage does the system cover?

   _____

   b. What is the daily rainfall equivalent (in inches) of the water from this system if 900,000 gallons of water are handled each day?

   _____

45. A homeowner has a 100-foot by 300-foot lot on which a 45-foot by 60-foot house stands. A driveway to the house is 18 feet wide and 120 feet long. The rest of the lot is lawn. What will it cost to apply turf fertilizer to the lawn if a $6.95 bag of fertilizer will cover 1,000 square feet? Assume that a partial bag may be purchased.

   _____

46. A frame enclosing a specific acreage of forage is used for research sampling. The forage inside the frame is cut, weighed, and tested for moisture content. If the sample is to represent 1/2,000 of an acre, what length should a 4-foot-wide frame have?

   _____

47. For a grain drill calibration test, field travel is simulated by jacking up one end of the drill and rotating the raised drive wheel as grain is collected from a specified number of planting units. Calculate the planting rate in bushels per acre from the following data:

   (a) drive wheel radius is 15 inches under load,
   (b) 1.5 pounds of barley collected from two units that are 7 inches apart,
   (c) 48 pounds of barley equals 1 bushel, and
   (d) drive wheel turned 65 revolutions.

   _____

48. The recommended planting rate for a certain variety of soybeans is 130,680 plants per acre. If the soybeans are planted in 20-inch rows, how many plants should there be per foot of row?

_____

49. Find the acreage in the field shown in Figure 10–14.

_____

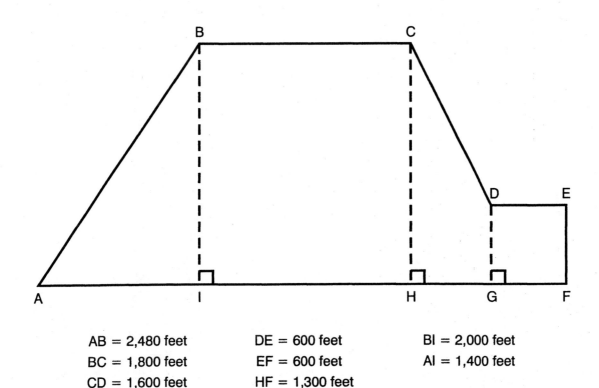

AB = 2,480 feet   DE = 600 feet   BI = 2,000 feet
BC = 1,800 feet   EF = 600 feet   AI = 1,400 feet
CD = 1,600 feet   HF = 1,300 feet

Figure 10–14

Exercise 10–3

ENGLISH AND METRIC VOLUME MEASUREMENTS AND FORMULAS

Refer to the Formulas of Mensuration (Appendix, Section 1) and Tables 7, 8, and 9 (Appendix, Section 2), English and Metric Volume Measurements and Conversions.

Add and simplify:

1. 4 gal 3 qt 1 pt + 6 gal 2 qt 1 pt + 1 gal 1 qt 1 pt   _____

2. 13 gal 1 pt + 3 qt 1 pt + 5 gal 1 qt   _____

Subtract:

3. 5 gal 2 qt – 3 gal 3 qt 1 pt   _____

4. 7 gal 2 qt – 5 gal 3 qt 1 pt   _____

Without measuring or converting units, estimate the correct response to the following statements using *kL, L,* or *mL.*

5. An auxiliary fuel tank for a semi would hold 95 _____ of fuel.

6. A veterinarian would inject 30 _____ of mastitis medication into a cow's udder.

7. A container used in a greenhouse for transplanting a single seedling would have a capacity of 150 _____.

8. An extra supply of gasoline for a riding lawn mower might have a capacity of 7.6 _____.

9. A large petroleum storage tank could have a capacity of 7,000 _____.

Make the following conversions:

10. _____ qt = 48 fl oz

11. 178 fl oz = _____ gal

12. 10.00 ft³ = _____ in.³

13. _____ cm³ = 45 mL

14. 15 m³ = _____ cm³

15. 55 cm³ = _____ in.³

16. 26.2 U.S. gal = _____ Imperial gal

17. 16.0 L = _____ qt

18. 96 pk = _____ bu

19. 10 gal = _____ pt

20. 565 in.³ = _____ gal

21. The owner of a fruit market purchased 18 bushels of apples. If the apples are to be sold in bags containing 1 peck 1 quart each, how many bags can be prepared for sale?

_____

22. A 12-fluid-ounce bottle of molasses retails for $1.19, while a 385-milliliter bottle of the same product retails for $1.25. Which size container of molasses is the better buy?

_____

23. How much wettable powder would you put into a 220-gallon tank if 11 gallons of water are applied per acre and 1.5 pounds of the commercial product are to be applied per acre?

_____

24. If an average oil well in the United States produces 15 barrels of oil per day, how many gallons of oil does the well produce per hour?

_____

25. Instructions for a 27-fluid-ounce can of grape juice concentrate state that the contents of the can should be diluted to make 1 gallon of grape juice. How many fluid ounces of water should be mixed with the concentrate?

_____

26. If leveling a building site for a new machine storage building required 3,400 cubic feet of fill, how many 14-cubic-yard truckloads of dirt were needed?

_____

27. Gasohol is selling for $1.41 per gallon. Find the cost of 25 liters of gasohol.

_____

28. An engine is quoted to have a 352-cubic-inch displacement. How many cubic centimeters of displacement does the engine have?

_____

29. A landscaping plan includes the placement of 108 cubic feet of pine bark between newly planted shrubs. If the bark can be purchased in 3-cubic-foot bags costing $3.05 each, what will the bark for the installation cost?

_____

30. A gas engine corn combine uses 2.32 gallons of gasoline per acre. Express this fuel requirement in liters per hectare.

_____

31. A gardener hired a backhoe operator to dig out the poor soil from a garden plot 24 feet by 18 feet to a depth of 15 inches. How many 5-cubic-yard truckloads of black dirt will be needed to fill the excavation?

_____

32. The hold of a grain barge is 47.2 meters long, 7.9 meters wide, and 2.1 meters deep. How many metric tons of rice can the barge carry if rice weighs 578 kilograms per cubic meter?

_____

33. A mechanized "hot walker" is to be installed at a riding stable. The level of the area where the installation is to be made needs to be raised 14 inches. The diameter of the installation is 40 feet. Because the fill material decreases in volume when it is packed, it is necessary to get approximately 20 percent more fill material than the packed fill needed. How many cubic yards of fill material will be required?

_____

34. The inside diameter of a cylindrical portion of a semen ampule is 5 millimeters. What is the length of the cylindrical portion of the ampule if it is designed to hold 0.7 milliliter of semen?

_____

35. A food product weighs 0.24 ounce per cubic inch. A manufacturer wants to package 38.4 ounces of the product in a carton that has a square base and is 10 inches tall. Find the needed dimensions for the carton's base.

_____

36. If 1 cubic foot of storage space holds 0.8 bushel, what diameter would an 18-foot-high, round steel bin have to be to store 8,200 bushels of grain sorghum?

_____

37. A concrete culvert has the dimensions shown in Figure 10–15. If a cubic foot of concrete weighs 140 pounds, find the weight of the culvert.

_____

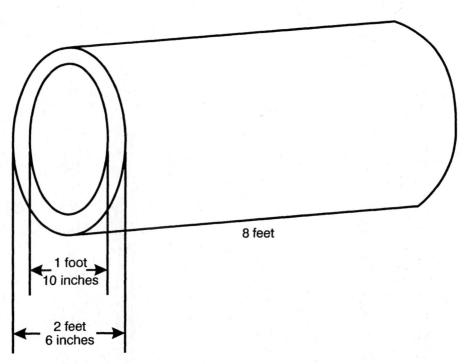

1 foot
10 inches

2 feet
6 inches

8 feet

Figure 10–15

38. A statement in an advertisement gives the capacity of a fuel storage tank as 560 gallons. If the cylindrical tank is 6 feet long and 4 feet in diameter, (a) is the actual capacity more or less than that given in the ad, and (b) by how much?

a. _____

b. _____

39. A commercial mixture of carbon disulfide and carbon tetrachloride can be purchased at $6.85 per gallon and is used to treat shelled corn for weevils at the rate of 3 gallons per 1,000 bushels. What would it cost to treat the shelled corn in a cylindrical bin 30 feet in diameter and 20 feet high? (One cubic foot of storage space holds 0.8 bushel of corn.)

_____

40. An outdoor feedrack for hay for young horses is 15 feet long, 5 feet deep, 6 feet 8 inches wide at the top, and 3 feet 4 inches wide at the bottom. How many bales of hay could be placed in the rack if the loose hay from a bale occupies approximately 12 cubic feet of space? (Assume that the rack is to be level full.)

_____

41. A sanitary boot bath is 7 inches high, 9 inches wide, and 18 inches long. It is desired to have the solution rise at least 5 inches on a single boot when immersed. If the average boot will displace approximately 250 cubic inches of liquid when immersed, how many gallons of sanitizing solution should be placed in one boot bath?

_____

42. A rectangular storage bin in an elevator is 28.6 feet long and 20.8 feet wide. To what depth is the bin filled if it contains 3,600 bushels of buckwheat, if one bushel of buckwheat occupies 1.25 cubic feet of space?

_____

43. A multi-dose veterinary syringe is designed to hold 60.0 milliliters of medication. If the length of the barrel of the syringe is 15.0 centimeters, what is the inside diameter of the barrel of the syringe?

_____

44. A packing carton has a square bottom 18 inches by 18 inches. If the capacity of the carton is 3 cubic feet, how deep is the carton?

_____

45. The family-sized can of orange juice is 9 inches tall and has a diameter of 5 inches. If there are approximately 58 cubic inches in a quart, how many 8-fluid-ounce servings are in the can?

_____

46. In order to properly design a heating system for a greenhouse, the volume of air in the building needs to be determined. Find the volume of air in the greenhouse shown in Figure 10–16.

_____

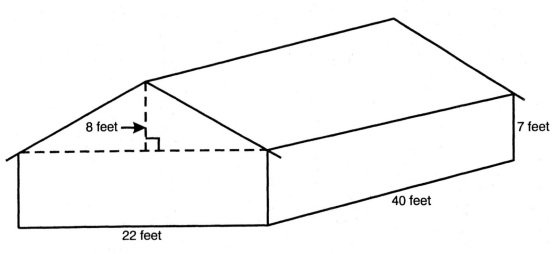

Figure 10–16

47. A greenhouse operator wants to fill six benches with soil that will have to be purchased. If each bench is 100 feet long and 4 feet wide, and will be filled to a depth of 6 inches, what will the soil cost at $10.50 per cubic yard?

_____

48. Floors of confinement hog units are often constructed from concrete slats similar to the one shown in Figure 10–17. How many of these slats can be made from a cubic yard of concrete? (Make no allowances for metal reinforcement.)

_____

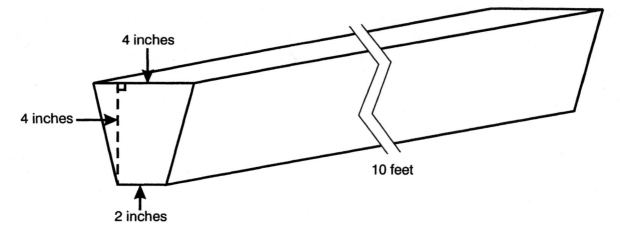

4 inches

4 inches —→

2 inches

10 feet

Figure 10–17

49. A stock tank with round ends has dimensions as shown in Figure 10–18. Find the number of kiloliters of water the tank would hold.

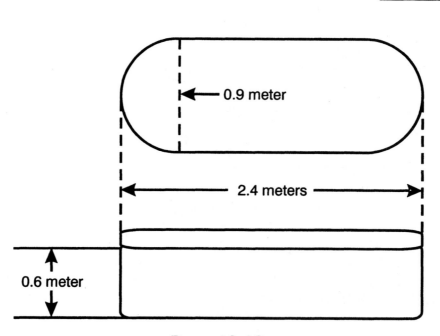

←— 0.9 meter

←——— 2.4 meters ———→

0.6 meter

Figure 10–18

50. The average annual precipitation in an area is 30.4 inches. At this rate, how many gallons of rainwater are deposited on each acre of land each year in this area?

Exercise 10–4

ENGLISH AND METRIC WEIGHT MEASUREMENTS

Refer to Tables 10, 11, and 12 (Appendix, Section 2), English and Metric Weight Measurements and Conversions; also Table 13 (Appendix, Section 2), Density and Specific Gravity.

Add and simplify:

1. 7 lb 12 oz + 2 lb 7 oz + 12 lb 9 oz   _____

2. 16 lb 2 oz + 3 lb 15 oz + 5 lb 8 oz   _____

Subtract:

3. 14 lb 4 oz – 6 lb 7 oz   _____

4. 25 lb 15 oz – 4 lb 8 oz   _____

Without measuring or making conversions, estimate the correct response to the following statements using *m.t.*, *kg*, or *g*.

5. A semi loaded with grain could weigh 23 _____.

6. A glass of milk might weigh 250 _____.

7. A beef steer ready for market weighs 0.5 _____.

8. A bushel of corn weighs 25.5 _____.

9. One soybean seed weighs 0.2 _____.

Make the following conversions:

10. 6.5 g = _____ mcg

11. _____ g = 6,780 cg

12. 2,860 lb = _____ m.t.

13. 75.0 kg = _____ lb

14. _____ cwt = 375 lb   15. 16.4 lb = _____ g

16. 225 oz = _____ lb   17. 2.54 T = _____ lb

18. 24.3 kg = _____ dag   19. 12 oz = _____ g

20. 5 g = _____ gr

21. If bulk milk is selling for $12.90 per hundredweight, what is the price per kilogram?

_____

22. A 26.25-ton load of grain is to be shipped by truck to a processing plant located 400 miles away. If the trucking rate for a 400-mile trip is $1.17 per hundredweight, what is the transportation cost?

_____

23. A family picked 75 pounds of berries at a pick-your-own strawberry plantation. If the price of pick-your-own berries is 35 cents per quart and a quart of berries weighs 1.25 pounds, what will the berries cost?

_____

24. What does a gallon of diesel fuel weigh?

_____

25. A grocer's ad quotes tom turkeys selling for 88 cents per pound. What is the cost of a turkey weighing 23 pounds 10 ounces?

_____

_____

26. A sprayer tank has a capacity of 560 gallons of water. What does a tankful of water weigh?

_____

27. In a principal commercial popcorn-producing state, the average yield is 3,450 pounds per acre. Convert this yield to kilograms per hectare.

_____

28. Anhydrous ammonia is 82 percent nitrogen and weighs 5.2 pounds per gallon. If a soil test recommends a 260-pound application of urea (46 percent nitrogen), how many gallons of anhydrous ammonia need to be applied to give an equivalent amount of nitrogen?

_____

29. Maple syrup is cooked until it has a specific gravity of 1.32. What does a pint of maple syrup weigh?

_____

30. A seed catalog lists a 1/128-ounce packet of carrot seeds containing 200 seeds selling for 95 cents.

a. When sold in this size packet, what is the value of 1 ounce of seed?

_____

b. How many seeds are in an ounce of seed?

_____

31. A sample of shelled corn was found to have 28 percent moisture. Upon drying to the moisture-free stage, the sample had lost 1.26 kilograms. What was the weight of the original sample?

_____

32. A comparison is being made of gross income from tobacco and peanuts. If the yield of peanuts is 2,643 pounds per acre and peanuts sell for 21.1 cents per pound, what yield in pounds per acre of tobacco selling at 117.4 cents per pound would give an equivalent income?

_____

33. A truck garden operation harvested 600 bushels of potatoes. The potatoes are graded and packaged in 15-pound bags and delivered to a local grocery store as needed. After 1,800 bags of potatoes had been delivered, how many bushels of potatoes remained to be processed? (One bushel of potatoes weighs 60 pounds.)

_____

34. A sweet corn grower had an average yield of 5 tons of sweet corn ears per acre from a 30-acre field. If the edible portion of the sweet corn is 24 percent of the harvested ears, how many boxes of frozen corn containing 20 ounces of corn could be packed from this field of corn?

_____

35. Commercial emulsifiable concentrate of methoxychlor has 2 pounds of active ingredient per gallon. If 0.78 pound of active ingredient is needed per acre, what amount of the concentrate should be put into a 50-gallon tank when the sprayer applies 10 gallons per acre?

_____

Name_____   Date_____

36. A concrete culvert contains 25.6 cubic feet of concrete. How many tons does the culvert weigh?

_____

37. A standing crop of timothy-legume hay is being offered for sale. If baled hay is selling for $72 per ton and harvesting costs are considered to be 55 percent of the total value of baled hay, what price should be charged per 65-pound bale harvested from the standing crop?

_____

38. A variety of cultivated wild rice yields 1,420 pounds per acre. If a grower had 75 acres of this variety and received a wholesale price of $5.10 per pound for the crop, what gross income did the grower have from wild rice?

_____

39. The oil extracted from soybeans amounts to 18.5 percent of the weight of the beans. If the specific gravity of soybean oil is 0.922, how many gallons of soybean oil can be made from a 60-pound bushel of soybeans?

_____

40. As a protein supplement for swine, linseed meal has 35.1 percent protein while cottonseed meal has 41.0 percent protein. If linseed meal is selling for $144.40 per ton and cottonseed meal is selling for $186.10 per ton, (a) which product is the more economical source of protein, and (b) what is the difference in price per pound of protein?

a. _____

b. _____

Exercise 10–5

TIME, TEMPERATURE, ANGULAR MEASUREMENT

Refer to Tables 14, 15, and 16 (Appendix, Section 2), Time, Temperature, and Angular Measurements.

Add and simplify:

1. 4 hr 48 min 36 sec + 7 hr 26 min 42 sec +
    9 hr 23 min 17 sec   _____

2. 2 week 4 da 16 hr + 1 week 20 hr + 1 week 6 da 15 hr   _____

Subtract:

3. 2 yr 2 mo 4 da – 11 mo 28 da   _____

4. 21 hr 45 min 6 sec – 16 hr 28 min 54 sec   _____

Without converting to Fahrenheit temperatures, estimate the best response to problems 5 through 9 by indicating A, B, C, D, or E.

A. 0°C to 2°C	B. 40.5°C	C. 22°C	D. –5°C	E. 10°C

5. Suitable temperature to start planting corn.   _____

6. Good temperature for apple storage.   _____

7. A hog with this temperature may need the attention of
    a veterinarian.   _____

8. The tractor radiator should have plenty of antifreeze at
    this temperature.   _____

9. This temperature is typical for a comfortable summer day.   _____

Make the following conversions:

10. 40 mo = _____ yr

11. 15 min = _____ hr

12. 113°F = _____ °C

13. 1307 (24-hour time) = _____ (a.m. or p.m. time)

14. 2¾ hr = _____ min

15. 18°F = _____ °C

16. 1 wk = _____ hr

17. −40°C = _____ °F

18. 9:14 p.m. = _____ (24-hour time)

19. 540 sec = _____ hr

20. 85°C = _____ °F

21. Much hay spoilage is due to the activity of heat-resistant fungi, which thrive when the temperature of the hay is from 113 degrees Fahrenheit to 150 degrees Fahrenheit. Find the range of Celsius temperatures at which these fungi are active.

_____

22. A custom combine operator harvested corn for a farmer from 7:30 a.m. until 5:45 p.m. If the rate is $80.60 per hour, what is the amount of the bill?

_____

23. The ideal temperature range for corn growth is at temperatures of 24 degrees to 30 degrees Celsius. What is this ideal temperature range on the Fahrenheit scale?

_____

_____

24. A thoroughbred horse ran a 1-mile race in 1 minute, 32 and $^1/_5$ seconds. What was the horse's speed in miles per hour?

_____

25. During a thunderstorm, the temperature dropped 15 Fahrenheit degrees. What was the temperature change in Celsius degrees?

_____

26. A standard speed for a power takeoff on a tractor is 540 revolutions per minute. What is the speed in revolutions per second?

_____

27. Spontaneous combustion of forages is the cause of many barn and silo fires. The ignition point of forages can be as low as 231 degrees Celsius. Give this ignition temperature for the Fahrenheit scale.

_____

28. If a tractor makes a 180-rod trip across a field in 5.6 minutes, (a) what is the tractor's speed in miles per hour, and (b) what is the speed in kilometers per hour?

_____

_____

Many seed companies provide information on the average number of **Growing Degree Days (GDD)** required for corn hybrids to reach maturity. The formula for calculating GDD is:

$$\frac{T.Max + T.Min}{2} - 50 = GDD$$

T.Max is the highest temperature during a day and T.Min is the lowest temperature. The GDD is calculated for each day and accumulated during the growing season. Because plant growth slows above 86°F and below 50°F, 86°F is substituted for T.Max when the temperature is above 86°F and 50°F is substituted for T.Min when the temperature goes below 50°F.

Use the Growing Degree Days Formula to solve problems 29 and 30.

29. It takes approximately 100 Growing Degree Days for corn to emerge regardless if it takes a few days or a few weeks. The day after a corn field is planted, the temperature ranged from 66 degrees Fahrenheit to 84 degrees Fahrenheit. How many days like this are needed for the corn to emerge?

_____

30. Recorded temperatures for five consecutive days in July were as follows:

	High	Low
July 2	91°F	74°F
July 3	83°F	67°F
July 4	68°F	48°F
July 5	75°F	67°F
July 6	82°F	68°F

a. Find the average daily Growing Degree Days during the period.

_____

b. If the growing season averages this number of daily Growing Degree Days, how many days will a corn hybrid listed as requiring 2,640 Growing Degree Days need for maturity?

_____

Exercise 10–6

ANGULAR MEASUREMENT

The most familiar unit of angular measurement is the **degree**, which is 1/360 of a revolution. The degree may be divided into **60 minutes**, and one minute may be sub-divided into **60 seconds**.

Angles may be added and subtracted just like other denominate numbers.

Example: Add 20° 45' 56" and 45° 25' 45".

$$
\begin{array}{lll}
20° & 45' & 56" \\
45° & 25' & 45" \\
\hline
65° & 70' & 101"
\end{array}
\qquad \text{or} \qquad
\begin{array}{lll}
65° & & \\
1° & 10' & \\
& 1' & 41" \\
\hline
66° & 11' & 41"
\end{array}
$$

Example: Subtract 26° 25' 45" from 45° 15' 10".

$$
\begin{array}{lll}
45° & 15' & 10" \\
26° & 25' & 45" \\
\hline
\end{array}
\qquad \text{or} \qquad
\begin{array}{lll}
44° & 74' & 70" \\
26° & 25' & 45" \\
\hline
18° & 49' & 25"
\end{array}
$$

For some mathematical operations it is often necessary to convert degrees and minutes to decimal degrees and vice versa.

Example: $67.7° = 67 + .7 \times 60 = 67° \ 42'$

Example: $29° \ 36' = 29 + \dfrac{36}{60}$ or $29.6°$

Add and simplify:

1. 17° 33' 26" + 42° 17' + 21° 2' 46" + 35° 24' 16"   _____

2. 29° 41' 16" + 4° 38' 52" + 20° 11' 48"   _____

Subtract:

3. 90° – 28° 45' 18" _____

4. 68° 4' 6" – 24° 28' 35" _____

Convert the following decimal degree angles to degrees and minutes:

5. 43.5° = _____

6. 145.3° = _____

7. 71.9° = _____

8. 23.45° = _____

9. 6.15° = _____

Convert the following to decimal degrees:

10. 132° 15' = _____

11. 5° 24' = _____

12. 62° 42' = _____

13. 43° 18' = _____

14. 89° 45' = _____

Name_____   Date_____

Use a protractor to construct the following angles. Use the • for the vertex of the angle.

15. 35° •————————————

16. 75° •————————————

17. 115° •————————————

18. 210° •————————————

19. Measure the degrees in each of the angles of the triangle below. Find the sum of the angles.

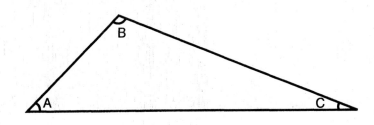

Angle A = _____

Angle B = _____

Angle C = _____

Total = _____

20. A hog farmer's records show that operating costs (feed costs not included) amount to $6.90 per hundredweight of pork produced. In preparing a circle graph to illustrate how the operating costs are distributed, it is found that each category should be represented by the following number of degrees:

Marketing and hauling	—	79°
Breeding and veterinary	—	65°
Electricity and fuel	—	29°
Grinding and mixing feed	—	43°
Equipment repair	—	29°
Interest and insurance	—	115°

Use the circle below to lay out the sector that represents each of the operating costs.

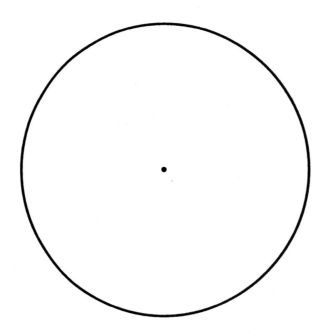

Appendix

Measurement Formulas

Measurement Formulas calculate the dimensions, areas, and volumes of geometric figures. In the formulas given below, the following letters are used consistently; others as indicated.

A = area   S = lateral surface area   h = altitude

B = area of prism base   C = circumference   d = diameter

K = total surface area   P = perimeter   r = radius

Perimeters and Areas

Square

$P = 4s$

$A = s^2$

Rectangle

$P = 2l + 2w$

$A = lw$

Circle

$C = 2\pi r$

$C = \pi d$

$A = \pi r^2$

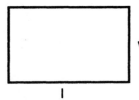

Parallelogram

$P = 2a + 2b$

$A = bh$

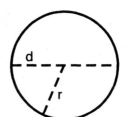

Trapezoid

$P = a + b_1 + b_2 + c$

$A = \frac{1}{2}(b_1 + b_2)h$

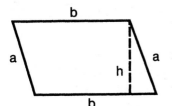

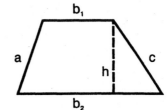

General Triangle

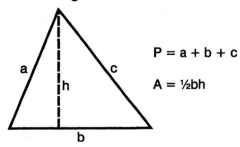

$P = a + b + c$

$A = \frac{1}{2}bh$

Right Triangle

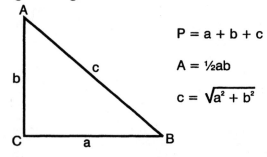

$P = a + b + c$

$A = \frac{1}{2}ab$

$c = \sqrt{a^2 + b^2}$

Volumes

Cube

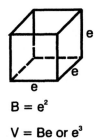

$B = e^2$

$V = Be \text{ or } e^3$

Rectangular Prism

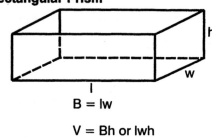

$B = lw$

$V = Bh \text{ or } lwh$

Triangular Prism

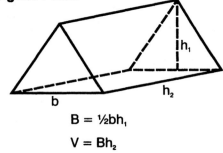

$B = \frac{1}{2}bh_1$

$V = Bh_2$

Trapezoidal Prism

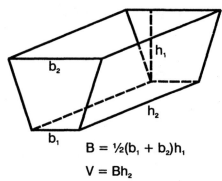

$B = \frac{1}{2}(b_1 + b_2)h_1$

$V = Bh_2$

Cylinder

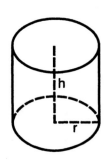

$S = 2\pi rh$

$K = S + 2(\pi r^2)$

$V = \pi r^2 h$

Cone

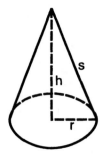

$S = \pi rs$

$K = S + \pi r^2$

$V = \frac{1}{3}\pi r^2 h$

Sphere

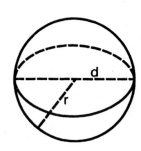

$K = 4\pi r^2 \text{ or } \pi d^2$

$V = \frac{4}{3}\pi r^3$

Appendix

English and Metric Measurements

Table 1

English Linear Units

12 inches (in. or ")	=	1 foot (ft or ')
3 feet	=	1 yard (yd)
16½ feet	=	1 rod (rd)
5½ feet	=	1 rod
320 rods	=	1 mile (mi)
5,280 feet	=	1 mile
1,760 yards	=	1 mile
1 furlong (fur)	=	1/8 mile or 660 feet

Table 2

Metric Linear Units

1 kilometer (km)	=	1,000 meters (m)
1 hectometer (hm)	=	100 meters
1 dekameter	=	10 meters
1 meter	=	1 meter
1 decimeter (dm) = 0.1 m		1 m = 10 dm
1 centimeter (cm) = 0.01 m		1 m = 100 cm
1 millimeter (mm) = 0.001 m		1 m = 1,000 mm

Table 3

English-Metric Conversion Units

1 in. = 25.4 mm	1 mm = 0.03937 in.
1 in. = 2.54 cm	1 cm = 0.3937 in.
1 ft = 30.48 cm	1 m = 39.37 in.
1 ft = 0.3048 m	1 m = 3.281 ft
1 yd = 0.9144 m	1 m = 1.0936 yd
1 rd = 5.029 m	1 km = 0.6214 mi
1 mi = 1.6093 km	
1 fur = 201.168 m	

Note: Use of alternate unit equations in Table 3 (e.g., 1 in. = 2.54 cm or 1 cm = 0.3937 in.) will result in slightly different answers.

Table 4

English Area Units

144 square inches (sq in. or in.^{2})	=	1 square foot (sq ft or ft^2)
9 square feet	=	1 square yard (sq yd or yd^2)
30¼ square yards	=	1 square rod (sq rd or rd^2)
160 square rods	=	1 acre (A)
43,560 square feet	=	1 acre
640 acres	=	1 square mile (sq mi or mi^2)

Table 5

Metric Area Units

1 square centimeter (cm^2)	=	100 square millimeters (mm^2)
1 square decimeter (dm^2)	=	100 square centimeters
1 square meter (m^2)	=	100 square decimeters
1 square meter	=	10,000 square centimeters
100 square meters	=	1 are (a)
100 ares	=	1 hectare (ha)
10,000 square meters	=	1 hectare

Table 6

English-Metric Conversion Units

1 in.^{2}	=	6.452 cm^2	1 mm^2	=	0.001549 in.^{2}
1 ft^2	=	0.0929 m^2	1 cm^2	=	0.1549 in.^{2}
1 yd^2	=	0.8361 m^2	1 m^2	=	1,549 in.^{2}
1 mi^2	=	259 ha	1 m^2	=	10.76 ft^2
1 mi^2	=	2.589 km^2	1 m^2	=	1.196 yd^2
1 rd^2	=	25.293 m^2	1 ha	=	2.471 A
1 A	=	40.47 a	1 a	=	119.6 yd^2
1 A	=	0.4047 ha			

Note: Use of alternate unit equations in Table 6 (e.g., 1 ft^2 = 0.0929 m^2 or 1 m^2 = 10.76 ft^2) will result in slightly different answers.

Table 7

English Cubic or Volume Measurements

Space or Solid Volume

1,728 cubic inches (cu in. or in.^{3}) = 1 cubic foot (cu ft or ft^3)

27 cubic feet = 1 cubic yard (cu yd or yd^3)

128 cubic feet = 1 cord (cd) — measure of wood

Liquid Measure

4 gills (gi) = 1 pint (pt)

1 pint = 16 fluid ounces (fl oz)

2 pints = 1 quart (qt)

4 quarts = 1 gallon (gal)

31½ gallons = 1 barrel (bbl)

7.48 gallons = 1 cubic foot (ft^3)

Dry Measure

2 pints (pt) = 1 quart (qt)

8 quarts = 1 peck (pk)

4 pecks = 1 bushel (bu)

1 bushel = 1.25 cubic feet (ft^3)

Household Measure

1 teaspoon (t) = $\frac{1}{6}$ fluid ounce (fl oz)

3 teaspoons = 1 tablespoon (T)

1 tablespoon = ½ fluid ounce

16 tablespoons = 1 cup (c)

1 cup = 8 fluid ounces

2 cups = 1 pint (pt)

2 pints = 1 quart (qt)

4 quarts = 1 gallon (gal)

Table 8

Metric Cubic or Volume Measurements

Space or Solid Volume

1 cubic centimeter (cm³) = 1,000 cubic millimeters (mm³)

1 cubic meter (m³) = 1,000,000 cubic centimeters

1,000 cubic centimeters = 1 cubic decimeter (dm³)

1,000 cubic centimeters = 1 liter (L)

1 cubic centimeter = 1 milliliter (mL)

Liquid Volume

1 kiloliter (kL) = 1,000 liters (L)

1 hectoliter (hL) = 100 liters

1 dekaliter (daL) = 10 liters

1 liter = 1 liter   1 liter = 1,000 cubic centimeters (cm³)

1 deciliter (dL) = 0.1 liter   1 liter = 10 deciliters

1 centiliter (cL) = 0.01 liter   1 liter = 100 centiliters

1 milliliter (mL) = 0.001 liter   1 liter = 1,000 milliliters

Table 9

English-Metric Conversion of Volume Measurements

Space or Solid Volume

1 in.³	= 16.387 cm³	1 cm³	= 0.06102 in.³
1 ft³	= 0.0283 m³	1 m³	= 35.32 ft³
1 yd³	= 0.7646 m³	1 m³	= 1.208 yd³

Liquid Measure

1 gi	= 7.219 in.³	1 gi	= 0.1183 L
1 pt	= 28.875 in.³	1 pt	= 0.4732 L
1 qt	= 57.75 in.³	1 qt	= 0.9463 L
		1 L	= 1.0567 qt
1 gal	= 231 in.³	1 gal	= 3.7853 L
		1 L	= 0.264 gal

Dry Measure

1 pt	= 33.6 in.³	1 pt	= 0.5506 L
1 qt	= 67.20 in.³	1 qt	= 1.1012 L
		1 L	= 0.908 dry qt
1 pk	= 573.61 in.³	1 pk	= 8.8096 L
1 bu	= 2,150.42 in.³	1 bu	= 35.2383 L
1 bbl	= 7,056 in.³	1 bbl	= 115.62 L

Imperial (Canada—Great Britain)

1 Imperial (Imp) qt	= 1.2009 United States (U.S.) qt
1 Imperial qt	= 69.3185 in.³
1 Imperial gal	= 1.201 United States gal
1 Imperial gal	= 277.42 in.³

Note: Use of alternate unit equations in Table 9 (e.g., 1 gal = 3.7853 L or 1 L = 0.264 gal) will result in slightly different answers.

Table 10

English Weight Units

Avoirdupois Weight

437.5 grains (gr)	=	1 ounce (oz)
7,000 grains	=	1 pound (lb)
16 ounces	=	1 pound
100 pounds	=	1 hundredweight (cwt)
2,000 pounds	=	1 ton (T) — short ton
2,240 pounds	=	1 long ton

Table 11

Metric Weight Units

1 metric ton (m.t.)	=	1,000 kilograms (kg)
1 kilogram	=	1,000 grams (g)
1 hectogram (hg)	=	100 grams
1 dekagram (dag)	=	10 grams
1 gram	=	1 gram
1 decigram (dg)	=	0.1 gram
1 centigram (cg)	=	0.01 gram
1 milligram (mg)	=	0.001 gram
1 microgram (mcg)	=	0.000001 gram

Table 12

English-Metric Conversion of Weight Measurements

1 gr	= 0.0648 g	1 g	= 15.432 gr
1 oz	= 28.35 g	1 g	= 0.03527 oz
1 lb	= 453.6 g		
1 lb	= 0.4536 kg	1 kg	= 2.2046 lb
1 T	= 0.9072 m.t.	1 m.t.	= 1.1023 T

Note: Use of alternate unit equations in Table 12 (e.g., 1 lb = 0.4536 kg or 1 kg = 2.2046 lb) will result in slightly different answers.

Table 13

Density — Specific Gravity

Density: The density of a substance is its weight (mass) per unit of volume.

Common statements of density:

One cubic foot of water weighs 62.4 pounds.

One liter of water weighs 1,000 grams.

One gallon of water weighs 8.34 pounds.

Specific gravity: The specific gravity of a substance is the number of times it is heavier than water. For example: The specific gravity of gasoline is 0.74. This means that a gallon of gasoline weighs 0.74×8.34 lb/gal or 6.17 pounds.

Common Specific Gravities	
Water	1.00
Gasoline	0.74
Ethyl alcohol	0.79
Diesel fuel	0.85
Concrete	2.3
Lead	11.3
Steel	7.8

Table 14

Time Units

60 seconds (sec)	=	1 minute (min)
60 minutes	=	1 hour (hr)
24 hours	=	1 day (da)
7 days	=	1 week (wk)
4 weeks	=	1 month (mo)
52 weeks	=	1 year (yr)
360 days	=	1 commercial year
30 days	=	1 commercial month

Table 15

Temperature Scales

Fahrenheit (F) Scale	*Celsius (C) Scale*
Freezing point = 32°	Freezing point = 0°
Boiling point = 212°	Boiling point = 100°
$F = \dfrac{9}{5}C + 32$	$C = \dfrac{5}{9}(F - 32)$

Growing Degree Days (GDD)

$$GDD = \frac{T.Max + T.Min}{2} - 50$$

Table 16

Angular Measurements

1 revolution or perigon	=	360 degrees (°)
1 degree	=	60 minutes (')
1 minute	=	60 seconds (")

Appendix

SECTION 3

Web Sites for Agricultural Information

National FFA Organization	www.ffa.org/
National Council for Agricultural Education	www.teamaged.org
National Association of Agricultural Educators	www.teamaged.org
Interstate Publishers, Inc.	www.IPPINC.com
U.S. Department of Education	www.ed.gov/
U.S. Department of Agriculture	www.usda.gov/
Carolina Biological Supply	www.carolina.com/
Freshwater Institute	www.conservationfund.org/
Livestock Breeds	www.ansi.okstate.edu/breeds/
AquaNIC	ag.ansc.purdue.edu/aquanic/
Hach Company	www.hach.com/
Stanley Tools	www.stanleyworks.com/
Case Corporation	www.casecorp.com
Monsanto	www.monsanto.com/
Farm Journal	www.FarmJournal.com
Agriculture Online	www.agriculture.com/
National Agricultural Statistics Service	www.usda.gov/nass/
USDA Economics and Statistics System	usda.mannlib.cornell.edu/usda/